BASIC Meta-Analysis: Procedures and Programs

Brian Mullen
Syracuse University

Robert Rosenthal
Harvard University

LAWRENCE ERLBAUM ASSOCIATES, PUBLISHERS
1985 Hillsdale, New Jersey London

Lawrence Erlbaum Associates, Inc., Publishers
365 Broadway
Hillsdale, New Jersey 07642

Library of Congress Cataloging in Publication Data

Mullen, Brian.
BASIC meta-analysis.

Bibliography: p.
Includes index.
1. Mathematical analysis—Data processing. 2. Basic (Computer program language) I. Rosenthal, Robert, 1933- . II. Title.
QA276.4.M85 1985 515′.028′5424 85-7049

ISBN 0-89859-702-1 (pbk.)

Typography by Dimensional Graphics, Roselle, New Jersey

Printed in the United States of America
10 9 8 7 6 5 4 3 2 1

BASIC Meta-Analysis: Procedures and Programs

To

Lou and Darcy

and

Jessica and Emily

Contents

Preface

This book presents an overview of meta-analytic techniques, and a series of BASIC computer programs that can be used to perform these meta-analytic procedures. Meta-analysis generally refers to the statistical integration of results of independent studies. This quantitative approach to the assessment of research domains represents a growing trend toward more rigorous and more precise summarization and integration of research literatures.

Chapter 1 presents a general framework for conceptualizing meta-analytic procedures, and describes the computational basis for each of fourteen meta-analytic techniques. Chapter 2 considers a number of practical aspects of applying these meta-analytic techniques, and presents general and specific decisions that must be made by anyone conducting a meta-analytic review. Chapter 3 discusses modifications of the BASIC computer programs that might be necessary or desirable for some computer systems. Chapter 4 uses a hypothetical set of studies to illustrate the use of each BASIC program to conduct the corresponding meta-analytic procedures. The illustration for each meta-analytic procedure presents the data that the program prompts the user to enter, the output displayed by the program, and a numerical illustration of the computation of the relevant meta-analytic statistics. Chapter 5 lists the BASIC meta-analysis programs. Chapter 6 presents the formulae and algorithms used to convert entered study data into the statistics used in the meta-analytic computations.

In presenting the conceptual basis of meta-analysis, this book may serve as a useful introduction to the quantitative assessment of research domains. Alternatively, this book presents practical guidelines for the application of meta-analytic techniques, and provides a package of computer programs which enhance the practicability of these techniques. Thereby, the book may also serve as a valuable addition for the experienced reviewer. The preparation we assume of our readers is that of an intermediate level course in quantitative methods (this preparation may be complemented by the reader's facility in entering, saving, and running programs on their own micro-, mini-, or mainframe computer systems).

It is our hope that this book will make meta-analytic techniques more accessible, and will encourage their responsible application.

The authors would like to express their appreciation to Mr. David P. Boynton, whose efforts and encouragement went above and beyond the call of duty. We would also like to thank Dr. Larry V. Hedges and Dr. Joseph S. Rossi for providing many helpful comments on earlier versions of this manuscript.

Further, the authors wish to express their gratitude respectively to Murray State University and Harvard University for support and encouragement.

Brian Mullen
Robert Rosenthal

When you can measure what you are speaking about, and express it in numbers, you know something about it; but when you cannot measure it, when you cannot express it in numbers, your knowledge is of a meager and unsatisfactory kind: it may be the beginning of knowledge, but you have scarcely, in your thoughts, advanced to the stage of *science*.

William Thomson,
Popular Lectures and Addresses (1891)

1 Meta-analysis

Meta-analytic statistics provide a single set of numbers that describe and summarize the results of independent pieces of research. Up until recently, attempts to review and integrate areas of research were confined to literary, narrative approaches. This so-called literary approach to review and integration of research has been criticized as being subjective, imprecise, and insensitive to specific details of the studies thereby reviewed. As Glass notes, "the accumulated findings of . . . studies should be regarded as complex data points, no more comprehensible without statistical analysis than hundreds of data points in a single study" (Glass, 1977, p. 352).

The seminal work of Glass and his colleagues (Glass, 1976, 1977, 1980; Glass, McGaw, & Smith, 1981; Smith & Glass, 1977; Smith, Glass, & Miller, 1980) is largely responsible for introducing meta-analysis to social scientists. Responses to this early work led to the development of a number of different meta-analytic techniques. As noted by Strube and Hartmann (1983), a meta-analytic review and summary of an area of research can serve many functions, such as: helping to direct future research more efficiently (Cooper, 1979); increasing the effectiveness with which policy decisions are made (Light, 1979); and, disseminating scientific information to wider audiences (Cooper & Rosenthal, 1980). Fiske (1983) further notes that meta-analysis requires researchers to: deal with the problems of fundamental categories and general definitions; exercise greater concern for the operationalization of the independent variable(s) and dependent variable(s); and, to conceptualize and theorize more sharply.

Meta-analytic techniques have recently been applied to a broad spectrum of phenomena in the social sciences. Consider the following partial list of phenomena which have been examined using meta-analytic techniques: interpersonal expectancy effects (Rosenthal, 1968; 1969; Rosenthal & Rubin, 1971, 1978); characteristics of volunteers for behavioral research (Rosenthal & Rosnow, 1975); the effectiveness of psychotherapy (Smith & Glass, 1977); characteristics of deceptive communication (DePaulo, Zuckerman, & Rosenthal, 1980; Zuckerman, DePaulo, & Rosenthal, 1981); effects of class size on attitudes and achievement (Glass & Smith, 1979); sex differences in social influence (Eagly & Carli, 1981); effects of public school desegregation (Israel, 1981); effects of computer assisted instruction (Bracey, 1982); validity of college student course ratings (Cohen, 1982); the effectiveness of tutoring (Cohen, Kulik, & Kulik, 1982); the efficacy of stimulant drug treatment for hyperactivity (Kavale, 1982); effects of ability grouping on student learning (Kulik & Kulik, 1982); cognitive gender differences (Rosenthal & Rubin, 1982c); the relation between socioeconomic status and academic achievement (White, 1982); the impact of leisure-time television on school learning (Williams, 1982); and, pretest sensitization effects in experimental design (Willson & Putnam, 1982).

In light of this large and growing currency received by meta-analysis, a number of reviews, critiques, and criticisms of meta-analysis have recently emerged (e.g., Cook & Leviton, 1980; Eysenck, 1978; Gallo, 1978; Glass et al. 1981; Leviton & Cook, 1981; Rachman & Wilson, 1980; Rosenthal, 1980; 1983a; Sohn, 1980; Strube & Hartmann, 1982; 1983). We concur with meta-analysis's critics who call for caution, responsible attention to detail, and explicit specification of the meta-analytic techniques employed. However, we are also encouraged by the tremendous potential of meta-analytic techniques to contribute to the needed production of useful cumulative knowledge (Cooper, 1982; Rosenthal, 1983b, in press; Schmidt, 1977). This volume represents an attempt to make the techniques of meta-analysis more accessible (and the conduct of meta-analyses more "time-efficient") by providing a series of BASIC meta-analysis programs for microcomputers.

A FRAMEWORK FOR META-ANALYTIC PROCEDURES

Within a given area of research, one study may report a product moment correlation coefficient based upon 100 subjects which supports the hypothesis under consideration; a second study may report a t-test based upon 50 subjects which also supports the hypothesis under consideration; and, another study may report a chi-square based on 300 subjects which shows no relationship or no effect. In a sense, meta-analysis is a method of statistical analysis wherein the units of analysis are the results of independent studies, rather than the responses of individual subjects.

Rosenthal (1983a & b) recently proposed a general framework for conceptualizing meta-analytic techniques. Table 1.1, based upon this framework, illustrates the fundamental attributes of the microcomputer programs provided in Chapter 5. Generally, there are two study outcomes that can be analyzed (significance levels and effect sizes), and there are two major analytic approaches (comparison of studies' results and combination of studies' results).

The *significance level* obtained for a single hypothesis test tells the researcher the possibility that the result obtained, or one more extreme, would have been obtained, if the null hypothesis were true. Probability levels (p levels) are reported in most published empirical studies which rely upon some type of inferential statistics. Meta-analytic procedures allow one to combine the p levels for a number of independent studies, in a number of ways. Note that imprecisely reported p levels (e.g., for a $t\ (30) = 3.03$, the p level might simply be reported as "$p<.05$") must be replaced by the most precise p levels which can be derived from the reported statistics (e.g., for a $t\ (30) = 3.03$, the corresponding p level is actually $p = .0025$). Extended tables of probability distributions or estimation algorithms (e.g.,

TABLE 1.1
Types of meta-analytic procedures, and corresponding BASIC programs (after Rosenthal, 1983a & b)

		Significance Levels		*Effect Sizes*
Two Studies	Comparison	A		B
	Combination	C		D
Three or More Studies	Comparison (Diffuse)	E		F
	Orthogonal Contrast Weights		G	
	Comparison (Focused)	H		I
	Combination	J		K
	Correlation Matrix, Partial Correlation, Multiple Correlation		L	
	Disjoint Cluster Analysis			M
	Model Estimation			N

Federighi, 1959; Abramowitz & Stegun, 1964; O'Grady, 1981; Rosenthal & Rosnow, 1984a) are helpful in this regard.

The *effect sizes* derived for a single hypothesis test tell the researcher the magnitude of the observed effects. The standardized difference between the means (*d;* Cohen, 1977) is the difference between group means, expressed on a standard normal metric ($d = (\bar{X}_1 - \bar{X}_2)/S.D.$). The larger the standardized difference between the means, the greater the magnitude of the effect. Another common index of effect size is the Pearson product moment correlation coefficient (r), and the associated proportion of variance accounted for (r^2). The larger the r and associated r^2, the greater the magnitude of the effect. Recently, Rosenthal and Rubin (1982a) have proposed an intuitively appealing, general-purpose display of effect size which is derived from the product moment correlation coefficient. This binomial effect size display (*BESD*) illustrates the effect of the independent variable of concern on the success rate/survival rate/improvement rate/etc. The *BESD* can be expressed as a change from one percentage or proportion to another percentage or proportion, and is defined as *BESD* = change from ($.50 - r/2$) to ($.50 + r/2$). Thus, an r of .30 results in a *BESD* of a change from 35% success rate/etc. to 65% success rate/etc. Meta-analytic procedures allow one to integrate the effect sizes for a number of independent studies, in a number of ways. Note that r's (and thereby r^2's and *BESD*'s) and d's can be derived from most commonly reported statistics (formulae given in Chapter 6).

The meta-analytic procedures of comparison indicate whether a set of studies has produced results (significance levels or effect sizes) that are significantly heterogenous. In other words, the meta-analytic tests of comparison reveal whether the studies are sufficiently different from one another to suggest that they should not be treated as coming from the same population. Sometimes, when there are three or more studies, theoretical concerns may lead one to expect systematic differences between the results of studies, as a function of methodological or practical aspects of the studies. In such cases, a focused test comparison can be conducted by weighing the study outcomes with theoretically-derived prediction or contrast weights (λ). For example, the contrast weights could be linear, quadratic, etc., coefficients of orthogonal polynomials. This type of focused test allows the researcher to conduct hypothesis testing at a level of research at which individual studies are the sampling units.

The meta-analytic procedures of combination provide a summary index of the overall results (significance levels or effect sizes) for a set of studies. In other words, the meta-analytic techniques of combination reveal the overall significance level or magnitude of effect in that area of research. Also, "fail safe numbers" can be derived that reveal how many new or unpublished studies showing no effect or null results would be needed to change the overall significance levels to some predetermined level.

As indicated in Table 1.1, the type of study outcome and the type of analytic approach, as well as the number of studies included in the meta-analysis, will define what type of meta-analytic procedure is appropriate. Each specific meta-analytic procedure, and a few supplementary procedures, are described in the following.

Meta-analytic Procedures for Two Studies

A. *Comparison of Significance Levels (Program A).* This meta-analytic procedure allows the reviewer to determine if the significance levels (p levels) of the two included studies differ significantly. From the statistics of each study, a reasonably exact one-tailed p value is derived. Note that one-tailed p values are always less than .50 when results are in the "expected direction" (i.e., when results support the hypothesis under consideration); however, one-tailed p values are always greater than .50 when results are in the "unexpected direction" (i.e., when results are inconsistent with the hypothesis under consideration). Reported two-tailed p values supporting the hypothesis under consideration should be halved; reported two-tailed p values inconsistent with the hypothesis under consideration should be halved and then subtracted from 1.00.

For each p value, the corresponding standard normal deviate (Z) is obtained. The Z's corresponding to one-tailed p's supporting the hypothesis will be positive, and the Z's corresponding to one-tailed p's inconsistent with the hypothesis will be negative. Alternatively, one can obtain the Z for significance level directly from most commonly reported statistics (formulae given in Chapter 6).

The formula (Rosenthal & Rubin, 1979) for the meta-analytic procedure of comparison of significance levels for 2 studies is:

$$Z = \frac{Z1 - Z2}{\sqrt{2}}$$

where: $Z1$ = Z associated with significance level of Study 1
$Z2$ = Z associated with significance level of Study 2

The p value corresponding to this new Z is the likelihood that this difference between the p values of the two studies, or a difference more extreme, could have occurred if the null hypothesis of no difference were true.

B. *Comparison of Effect Sizes (Program B).* This meta-analytic procedure allows the reviewer to determine if the effect sizes of the two included studies differ significantly. From the statistics of each study, effect sizes (r, r^2, d and the effect size display *BESD*) are derived. Zs are obtained for each study's effect sizes by performing Fisher's transformation on the rs for each study. Fisher's Z is defined as $Z = \frac{1}{2}\log_e\left(\frac{1+r}{1-r}\right)$. The formula (see Snedecor &

Cochran, 1967; 1980) for the meta-analytic procedure of comparison of effect sizes for two studies is:

$$Z = \frac{Z1 - Z2}{\sqrt{\frac{1}{N1-3} + \frac{1}{N2-3}}}$$

where: $Z1$ = Fisher Z associated with effect size for Study 1
$Z2$ = Fisher Z associated with effect size for Study 2
$N1$ = N associated with Study 1
$N2$ = N associated with Study 2

The p value corresponding to this new Z is the likelihood that a difference between the effect sizes of the two studies, or one more extreme, could have occurred if the null hypothesis were true.

C. *Combination of Significance Levels (Program C).* This meta-analytic procedure allows the meta-analyst to estimate the probability that the p levels of the two included studies might have been obtained if the null hypothesis were true. From the statistics of each study, a Z is obtained for the significance level. Each Z is then multiplied by a weight. The reviewer can use a weight of "1" (this would produce an unweighted Stouffer's [1949] method of combining significance levels). Alternatively, the reviewer could weight each Z by the relevant degrees of freedom, by the relevant sample sizes, by some index of the internal validity of the studies, etc. (this would produce a weighted Stouffer's [1949] method of combining significance levels). The formula for the meta-analytic procedure of combination of significance levels for two studies is:

$$Z = \frac{w1Z1 + w2Z2}{\sqrt{w1^2 + w2^2}}$$

where: w1 = weight assigned to results of Study 1
w2 = weight assigned to results of Study 2
$Z1$ = Z associated with significance level of Study 1
$Z2$ = Z associated with significance level of Study 2

The p value corresponding to this new Z is the likelihood that the combined results of the two studies, or one more extreme, would have been obtained if the null hypothesis were true.

D. *Combination of Effect Sizes (Program D).* This meta-analytic procedure allows the reviewer to estimate the mean effect size for the two included studies. From the statistics of each study, effect sizes (r, r^2,d and the effect size display *BESD*) are derived. Zs are obtained for each study's effect sizes by performing Fisher's transformation on the *r*s for each study. Each Fisher's *Z* is multiplied by a weight (e.g., "1", sample size, etc.). The formula for the meta-analytic procedure of combination of effect sizes for two studies is:

$$Z = \frac{w1Z1 + w2Z2}{w1 + w2}$$

where: w1 = weight assigned to results of Study 1
w2 = weight assigned to results of Study 2
Z1 = Fisher Z associated with effect size for Study 1
Z2 = Fisher Z associated with effect size for Study 2

The effect sizes (r, r^2, d and the effect size display *BESD*) associated with this new *Z* represent the combined effect size for the two studies.

Meta-analytic Procedures for Three or More Studies

E. *Comparison of Significance Levels (Diffuse Test) (Program E).* This procedure, an extension of the procedure described in *A* above, allows the reviewer to determine if the *p* levels of three or more included studies differ significantly. From the statistics of each study, a *Z* is obtained for the significance level. The formula (see Rosenthal & Rubin, 1979) for the meta-analytic procedure of comparison of significance levels (Diffuse test) for three or more studies is:

$$\chi^2_{(k-1)} = \Sigma\,(Zj - \bar{Z})^2$$

where: Zj = Z associated with significance level of Study j
$\bar{Z}$ = Mean Z
k = Number of studies

The *p* value corresponding to this χ^2 with *k*-1 degrees of freedom is the likelihood that a χ^2 as large or larger would have been obtained if the null hypothesis were true.

F. *Comparison of Effect Sizes (Diffuse Test) (Program F).* This procedure, an extension of the procedure described in *B* above, allows the meta-analyst to determine if the effect sizes of three or more included studies differ significantly. From the statistics for each study, effect sizes (r, r^2, d and the *BESD*) are derived. *Z*s are obtained for each study's effect sizes by performing Fisher's transformation on the *r*s for each study. The formula (see Snedecor & Cochran, 1967; 1980) for the meta-analytic procedure of comparison of effect sizes (Diffuse test) for three or more studies is:

$$\chi^2_{(k-1)} = \Sigma\,(Nj-3)\,(Zj-\bar{Z})^2$$

where: Zj = Fisher Z associated with effect size of Study j
$\bar{Z}$ = Weighted Mean Fisher Z
Nj = N associated with Study j
k = Number of studies

The p level corresponding to this χ^2 with k-1 degrees freedom is the likelihood that a χ^2 this large or larger would have been obtained if the null hypothesis were true.

G. *Linear and Quadratic Orthogonal Contrast Weights (Program G).* This program computes linear and quadratic orthogonal contrast weights for use in Programs H and I (i.e., for use in Focused test comparisons of significance levels and effect sizes, respectively). Using the procedure described in Kirk (1982, Appendix C), orthogonal polynomial coefficients for linear and for quadratic functions can be obtained for equal or unequal intervals of the attribute of concern, and for equal or unequal numbers of exemplars at each level of the attribute of concern.

H. *Comparison of Significance Levels (Focused test) (Program H).* The purpose of this focused comparison is to determine if the p values of three or more included studies are significantly predictable in some systematic, specifiable manner. From the statistics of each study, a Z is obtained for the significance level. Each Z is then multiplied by a contrast weight (λ) (e.g., linear or quadratic contrast weights derived from Program G). The formula (see Rosenthal & Rubin, 1979) for the meta-analytic procedure of comparison of significance levels (Focused Test) for three or more studies is:

$$Z = \frac{\Sigma\lambda j\, Zj}{\sqrt{\Sigma\lambda j^2}}$$

where: Zj = Z associated with significance level for Study j
λj = contrast weight assigned to results of Study j

The p level corresponding to this new Z reveals the likelihood that a Z as large or larger would have been obtained if the null hypothesis of no relationship between the predicted and obtained results were true.

I. *Comparison of Effect Sizes (Focused Test) (Program I).* The purpose of this focused comparison is to determine if the effect sizes of three or more included studies are significantly predictable in some systematic, specifiable manner. From the statistics of each study, effect sizes (r, r^2, d and the *BESD*) are derived. *Z*s are obtained for each study's effect sizes by performing Fisher's transformation on the *r*s for each study. Each Fisher's Z is multiplied by a contrast weight (λ) (e.g., linear or quadratic contrast weights derived from Program G). The formula (see Rosenthal & Rubin, 1982b) for the meta-analytic procedure of comparison of effect sizes (Focused Test) for three or more studies is:

$$Z = \frac{\Sigma \lambda j \, Zj}{\sqrt{\sum \frac{(\lambda j^2)}{(Nj - 3)}}}$$

where: Zj = Fisher Z associated with effect size for Study j
λj = contrast weight assigned to results of Study j
Nj = N associated with Study j

The p level corresponding to this new Z reveals the likelihood that a Z as large or larger would have been obtained if the null hypothesis of no relationship between the predicted and obtained results were true.

J. *Combination of Significance Levels (Program J).* This procedure, an extension of the procedure described in *C* above, allows the reviewer to estimate the probability that the p levels of three or more included studies might have been obtained if the null hypothesis were true. From the statistics of each study, a Z is obtained for the significance level. Each Z is then multiplied by a weight (e.g., "1", sample size, etc.). The formula for the meta-analytic procedure of combining significance levels for three or more studies is:

$$Z = \frac{\Sigma wjZj}{\sqrt{\Sigma wj^2}}$$

where: wj = weight assigned to results of Study j
Zj = Z associated with significance level of Study j

The p level corresponding to this new Z is the likelihood that a Z this large or larger would have been obtained if the null hypothesis were true.

Another aspect of the combination of significance levels involves the calculation of the number of unretrieved studies averaging null results needed to bring the overall combined probability level to some specified level. The formula (see Rosenthal, 1979) for this number for the $p = .05$ level of significance is:

$$N_{(p=.05)} = \left(\frac{\Sigma Zj}{1.645}\right)^2 - k$$

where: Zj = Z associated with significance level of Study j
k = Number of retrieved studies

This estimates the number of new, filed, or unretrieved studies averaging null results that would be required to bring the overall combined probability level to the "just significant" convention of $p = .05$.

K. *Combination of Effect Sizes (Program K).* This procedure, an extension of the procedure described in *D* above, allows the meta-analyst to estimate the mean effect size for the included set of studies. From the statistics of each study, effect sizes (*r*, r^2, *d* and the *BESD*) are derived. *Z*s are obtained for each study's effect sizes by performing Fisher's transformation of the *r*s for each study. Each Fisher's *Z* is then multiplied by a weight (e.g., "1", sample size, etc.). The formula for the meta-analytic procedure of combination of effect sizes for three or more studies is:

$$Z = \frac{\Sigma wjZj}{\Sigma wj}$$

where: wj = weight assigned to results of Study j
Zj = Fisher *Z* associated with effect size for Study j

The effect sizes (*r*, r^2, *d* and the *BESD*) corresponding to this new *Z* represent the combined effect sizes for the included studies.

L. *Correlation Matrix, Partial Correlation and Multiple Correlation (Program L).* This procedure is not inherently meta-analytic in nature, although it may be critically important in interpreting the results of a meta-analysis, or otherwise furthering one's understanding of a body of research. The focused test comparisons (described in sections *H* and *I* preceding) allow the meta-analyst to determine if a given study characteristic appears to influence study outcomes. The meta-analytic reviewer might also be interested in examining the interactive influence of many study characteristics on study outcomes. This procedure produces a matrix of Pearson product moment correlation coefficients; this can be useful in allowing the reviewer to examine, at a glance, the nature and the extent of bivariate relationships between study attributes and study outcomes. Further, the reviewer may examine partial correlations involving any three variables represented in the obtained correlation matrix. For example, the reviewer might wish to examine the correlation between effect size and level of independent variable manipulation, partialing out the influence of date of publication. Finally, the reviewer may examine multiple correlations involving any three variables represented in the obtained correlation matrix. For example, it may be informative to examine the multiple correlation between effect size as the "criterion" variable, and level of independent variable manipulation and proportion of the sample that is male as the "predictor" variables. These multivariate indices (Marascuilo & Levin, 1983) provide an approach to the examination of patterns of study outcomes that may be a useful complement to the more powerful Focused Test comparisons described earlier. If the meta-analysis is based upon a reasonably large number of retrieved studies, the results of these correlational procedures might reveal multivariate relationships that could then be examined using one of the many BASIC multiple regression programs currently available (cf. Madron, 1981; Chereb, 1979).

M. *Disjoint Cluster Analysis of Effect Sizes (Program M).* The procedures described in sections *H, I,* and *L* above are based upon the assumption that the reviewer has some a priori expectations regarding the variables that might influence study outcomes. That is, variables suspected to influence study outcomes are used to derive contrast weights (using Program G) for Focused Test comparisons (using Programs H and I), or, are directly entered into correlational procedures (Program L). However, studies may produce differing results for reasons not immediately apparent to the reviewer. For example, the studies included in a meta-analysis may have produced significantly heterogeneous effect sizes (as determined by a significant Diffuse Test comparison of effect sizes produced by Program F). However, the effect sizes of these studies are not significantly predictable in terms of contrast weights derived from any of the study attributes that the reviewer has examined (as determined by nonsignificant Focused Test comparisons of effect sizes produced by Program I). Further, the correlations, partial correlations, and multiple correlations (produced by Program L) also fail to suggest any meaningful explanation for the observed differences.

Hedges and Olkin (1983) describe a disjoint cluster analysis procedure which may be useful in this type of situation. The procedure identifies "disjoint," or non-overlapping, clusters of studies, based upon the effect size (r) and the sample size entered for each study (cf. Gibbons, Olkin, & Sobel, 1977). More specifically, all of the entered rs are rank ordered from highest to lowest. Then, each r is transformed into a Fisher's Z. Each Fisher's Z is then weighted by a square-mean-root estimate of common sample size to yield a standard normal variate (U):

$$Uj = \left(\frac{\Sigma \sqrt{Nj-3}}{k} \right) Zj$$

where: Nj = N associated with Study j
Zj = Fisher Z associated with effect size for Study j
k = Number of studies

The difference between adjacently-ranked Us are then examined. If two adjacently ranked Us are significantly different, then the corresponding two studies fall into different clusters. According to this approach, significance of the difference between Us is dependent upon the number of studies included in the cluster analysis. Hedges and Olkin (1983, Table 1) provide a table of critical values for determining whether a difference between the effect sizes of adjacently-ranked studies is significant, given the number of studies involved and given a specified level of statistical significance. This program utilizes linear approximations to these critical values (formulae provided in Chapter 6). If the difference between two adjacently ranked Us is equal to or exceeds the critical value for a selected level of statistical significance, the corresponding two studies are placed in different clusters; otherwise, those two studies are placed in the same cluster.

Once disjoint clusters have been identified, the reviewer may be able to work "backwards" from these clusters to some underlying explanation. The clusters of studies may be found a posteriori to differ on some previously unexamined study attribute. This study attribute could then be used to derive contrast weights (using Program G), which could in turn be used to conduct a new Focused Test comparison of effect sizes (using program I). As with any unplanned contrast, significance levels associated with these comparisons should be conservatively interpreted (Rosenthal & Rosnow, 1984b).

N. *Model Estimation (Program N).* This final procedure is based upon Glass's (1980; Glass & Smith, 1979) approach to meta-analytic model estimation. This approach to the integration of research provides an aggregate description of the relationship between two quantitative variables (the independent variable and the dependent variable in the area of research being summarized). For example, Glass (1980) described the estimation of a model which characterizes the relationship between class size and student achievement.

To begin this model estimation procedure, the reviewer derives an effect size (e.g., *d*) for all retrieved comparisons between larger and smaller exemplars of the quantitative independent variable; these effect sizes can be derived by entering the relevant study statistics into one of the effect size programs (i.e., Programs F, I or K). For example, the reviewer may derive a *d* for *k* comparisons between the achievement scores in a class of larger size and the achievement scores in a class of smaller size. The reviewer then specifies the nature of the model to be estimated. This specification takes the form of choosing the arithmetic operation to be performed on the independent variable values of the two exemplars for each comparison. The reviewer can specify the model to take the form of: the ratio of the larger exemplar to the smaller exemplar; the natural logarithm of the ratio of the larger exemplar to the smaller exemplar; or, any user-defined exponentiation of the ratio of the larger exemplar to the smaller exemplar.

This procedure then performs a regression analysis to test the accuracy of the specified model. That is, the effect size for each comparison is regressed upon the predictor formed by the model which was specified (cf., Kerlinger & Pedhazur, 1973). For example, Glass (1980) regressed the effect size for each comparison of achievement scores upon the logarithm of the ratio of the larger class size to the smaller class size. The results of this regression analysis indicate the ability of the specified model to account for variability in the effect sizes. The formula for the least-squares regression line provides the arithmetic estimate of the model that represents the relationship between the independent variable and the dependent variable. Note that the value of this approach does *not* lie in testing the significance of an independent variable's predictive utility. Rather, if a variable has previously been determined to be a statistically significant predictor of affect sizes (using the focused test comparison of effect sizes produced by Program I), Program N allows the researcher to develop a visual illustration of this relationship.

Each of these meta-analytic procedures is illustrated with specific examples in Chapter 4.

2 Applications of Meta-analytic Procedures

Cooper (1982, in press), Glass, McGaw, and Smith (1981), Hunter, Schmidt, and Jackson (1982), Rosenthal (1980, 1983a & b, in press) and Strube and Hartmann (1983) have provided useful descriptions of the issues and problems that are important in the application of meta-analytic techniques. This chapter draws heavily from, and attempts to summarize the basic points of, these sources; the serious meta-analyst should examine these more detailed presentations of the points summarized in this chapter.

Strube and Hartmann noted that "no amount of precision or quantification in the actual techniques can spare the reviewer the thought and planning that are required of *any* review" (Strube & Hartmann, 1983, p. 17). The nature of meta-analysis forces the user of these techniques to make a number of carefully considered decisions. Some critics (e.g., Wilson & Rachman, 1983) have argued that meta-analysis requires the researcher to make a number of subjective judgements, and this in turn vitiates any improvements that meta-analysts might purport to make upon traditional literary reviews. The prognosis for successful application of meta-analysis is not as dismal as this suggests. As Mintz (1983) points out, the narrative literary reviewer tends to use private subjective rules in the integration of a body of research; the meta-analyst is forced to make these subjective rules public and explicit. Nonetheless, it must be recognized that the quality and the value of a meta-analysis will depend upon the care and the rigor with which the techniques are applied.

With this in mind, it is advantageous to highlight some practical aspects of applying the meta-analytic techniques outlined in the previous chapter.

The approach to be taken here is to list a number of general and specific decisions which the meta-analytic reviewer must make. Our purpose is to help the users of meta-analysis to make explicit and public the steps underlying their application of meta-analytic procedures.

Decision 1. Procuring Studies. The first set of decisions to be made regards the process of procuring and selecting the studies to be included in the meta-analysis. This decision is in some ways the most critical; as Rachman and Wilson (1980) suggest, the results of a meta-analysis can be useless and even counterproductive if the meta-analysis is conducted upon an incomplete and biased data base. Three specific decisions are required in the process of procuring studies for a meta-analysis.

Decision 1a. Source of Studies. The user has to identify the sources of studies which will be included in the meta-analysis. Studies can be obtained from a variety of (overlapping) categories: articles published in professional journals; books; theses and dissertations; papers presented at regional and national conventions; archives; unpublished sources (e.g., unpublished manuscripts, manuscripts under editorial review).

Some meta-analysts (e.g., Cooper, 1979) rely upon studies obtained from published sources. This reliance may be based upon the assumed higher quality of published research, and certainly limits one's data base to studies that are most easily available. However, some evidence indicates that publication policies may be biased in favor of reporting significant findings (e.g., Sterling, 1959; Smart, 1964; Bakan, 1967; Rosenthal, 1966; Greenwald, 1975); hence, strict reliance on published sources could produce inflated estimates of the overall levels of significance and/or overall effect sizes. However, Rosenthal (in press) compared the effect sizes obtained from published studies, dissertations and theses, books, and unpublished sources, and found that published studies yielded effect sizes of about the same size as did unpublished studies and books. Dissertations and theses, however, did tend to yield somewhat smaller effect sizes than the other three sources. Utilizing the full range of sources of studies will produce a more comprehensive and more accurate meta-analysis.

The research attention devoted to some topics may present the reviewer with the mixed blessing of an extremely large number of studies. One approach to this situation is to select (as best as possible) a random, representative sample from this population of studies, and conduct the meta-analysis on this sample (e.g., Rosenthal & Rubin, 1978; and, Miller, reported in Glass et al. 1981). Another approach is to sharpen one's conceptual focus and perform separate meta-analyses on smaller samples of more closely related studies (e.g., Hall, 1980). Regardless of the approach taken, the reviewer must make the decision explicit and justify it.

Decision 1b. Locating Studies. The actual process of locating studies in a given area of research can be both tedious and time-consuming. Cooper (1982) has identified a number of techniques for locating studies. The *ancestry* approach retrieves information by following the bibliographies of relevant studies to locate earlier relevant studies. The *descendency* approach retrieves information by using indexing sources (e.g., *Social Science Citation Index*) to locate subsequent studies that have cited earlier relevant studies. The reviewer can use *abstracting services* (e.g., *Psychological Abstracts, Sociological Abstracts*) to locate studies associated with key words or phrases. *On-line computer searches* (e.g., ERIC, BIOSIS, PsycINFO; cf., Atkinson & Dolan, 1983) scan citation indexes and abstracting services at great speed. The *invisible college* approach describes accessing new and/or unpublished studies by corresponding with researchers who are active in the area of interest. Proceedings from professional association conventions may provide access to otherwise unpublished research.

Decision 1c. The "File Drawer" Problem. This refers to the possibility that unknown, unpublished studies might exist, whose results fail to support the pattern established by published findings. Rosenthal has previously stated the extreme view of the file drawer problem as follows: " . . . the journals are filled with the 5% of the studies that show Type I errors, while the file drawers back at the lab are filled with the 95% of the studies that show nonsignificant ($p > .05$) results" (Rosenthal, 1979, p. 638).

"Fail safe" numbers have been developed to indicate the number of new, filed or unretrieved studies which would be required to "undo" the results of a meta- analysis. Program J above produces a fail safe number for the .05 p value. This fail safe number estimates the number of unretrieved studies averaging null results that would be needed to bring the overall probability level to the "just significant" convention of $p = .05$. Applying this approach to a meta-analysis of 345 studies of the effects of interpersonal self-fulfilling prophecies, Rosenthal and Rubin (1978) determined that over 65,000 unreported studies averaging null results would have to exist somewhere before the overall results could be reasonably ascribed to sampling bias.

The meta-analytic reviewer should recognize that these fail safe numbers are based upon the values obtained from the studies included in the meta-analysis. If these studies are unrepresentative of the general population of studies examining the phenomenon in question, these fail safe numbers should be viewed with caution (cf., Strube & Hartmann, 1983).

Decision 2. Coding of Study Characteristics. The second set of decisions to be made regards the coding of study characteristics. After the studies to be included in the meta-analysis have been selected, it may be valuable to select a variety of study characteristics that may affect study outcomes. Two specific decisions are required in the coding of study characteristics.

Decision 2a. Choice of Study Characteristics. The phenomenon of concern will determine what types of study attributes should be coded. Quantitative attributes include such things as: date of publication; age of subjects; percentage of sample that is male/female; experimental mortality; and sample size. Some attributes may be amenable to straightforward categorization, such as: assignment of subjects (random vs. nonrandom); location of study (laboratory vs. nonlaboratory); publication form (journal, book, dissertation, etc.). Finally, many attributes will have to subjected to some quantifying coding scheme, such as: the internal validity of the study; the similarity between subjects and experimenters; the level of intensity of the independent variable manipulations; the reactivity of the dependent variable measurements; and, the overall quality of the study.

Some authors (e.g., Wilson & Rachman, 1983; Strube & Hartmann, 1982) have criticized the subjectivity of the decisions required to develop quantitative coding schemes. This criticism is well taken; for example, Gottfredson (1978) demonstrated relatively low agreement among professional psychologists in evaluating the quality of research. The most useful recommendation is that the meta-analyst should make explicit the rationale behind, and the procedure underlying, those coding schemes used. According to Cooper (1982), more than one rater should code these attributes, and indices of interrater agreement should be reported. Stock, Okun, Haring, Miller, Kinney, and Ceurvorst (1982) reported that the reliability of judgements of the quality of studies included in a meta-analysis may be improved by: using multiple dimensions of study quality; having descriptions of the dimensions printed on the codebooks used in recording study information; and, by holding training and practice sessions for all coders before and during the interval when study information is being recorded. For more information on the quality of studies and on the coding of this type of data, the reader is referred to Campbell and Stanley (1963), Webb, Campbell, Schwarz, and Sechrest (1966), and Olson (1976, Chapter 7) for general presentations, and to Glass et al. (1981), Smith et al. (1980), Rosenthal (1980, in press), and Stock et al. (1982) for presentations directed toward meta-analysis.

Decision 2b. Use of Study Characteristics. The study characteristics included in a meta-analysis can be used to suggest which variables (if any) may mediate reported significance levels and effect sizes. As described above, Programs H (for significance levels) and I (for effect sizes) can be used to determine if study outcomes vary in a predictable way which can be specified using contrast weights (such as those which can be obtained using Program G). Program L can be used to examine the interrelationships between study outcomes and each of a number of study attributes, as well as to examine potentially valuable partial correlation coefficients and multiple correlation coefficients.

The use of study characteristics in a meta-analysis can reveal patterns that would remain unnoticed by a literary reviewer. For example, Smith and Glass (1977) found that effect size for psychotherapy outcome was significantly correlated with the reactivity of the outcome measure, and with the IQ of the clients.

However, the use of study characteristics in a meta-analysis should be undertaken with caution. For one thing, applications of inferential statistical techniques (e.g., correlation, as in Program L; or, multiple regression) may not be strictly appropriate for meta-analytic statistics. Strube and Hartmann (1983) note that if the units of a meta-analysis (i.e., the significance levels and the effect sizes of the included studies) are not randomly and independently sampled from a known population, then the results of inferential statistical techniques using study characteristics should be viewed as only suggestive. Further, Cooper (1982, in press) cautions against inferring causality from any relationships observed between study characteristics and study outcomes. The nonspuriousness of such relationships observed in meta-analysis is always problematic. If a novel, theoretically relevant relationship is suggested by the examination of relationships between study characteristics and study outcomes, the reviewer should call for the relationship to be investigated at the level of experimental studies where spurious influences can be better controlled (Rosenthal, in press).

Decision 3. Data Retrieval. We hope that the growing popularity of meta-analytic procedures will encourage an increase in the precision of statistical reporting in future publications. At the very least, the original test statistic, an accurate p value corresponding to this test statistic, and the relevant sample size/degrees of freedom should be provided (if for no other reason than to allow posterity to most accurately represent the results of a given study in future meta-analyses). However, the meta-analytic reviewer must often make decisions that will affect the statistics for a given study included in the meta-analysis. Three specific decisions are considered in the following.

Decision 3a. Data Reconstruction. There are established techniques for deriving Z, r, and d from study statistics (cf. Glass et al. 1981; Rosenthal, 1978, 1980, in press). If a study fails to provide a complete report of statistical information, the meta-analytic reviewer may have to reconstruct the data necessary for a meta-analysis. Obviously, it would be preferable to obtain the original test statistics from the author(s) of incompletely reported studies included in a meta-analysis. However, if this is not possible, conservative estimates of study outcomes can be made. For example, descriptions of "no significant difference" in the absence of statistics can be conservatively interpreted as $p = .50$ (or $Z = 0.00$). Descriptions of "significant difference" in the absence of statistics can be conservatively interpreted as $p = .05$ (or $Z = 1.96$ if two-tailed *p*s can be assumed). Although these reconstructions are likely to be patently false in any given instance, they are the most defensible approximations available. The meta-analytic reviewer should report the number and the nature of those reconstructions which were necessary.

Decision 3b. Multiple Dependent Variable Measurements. As with similar statistical procedures employed at the level of primary analysis, the meta-analytic procedures described in the foregoing assume independence of the units of analysis (in the case of meta-analysis, independence of the individual hypothesis tests included in the meta-analysis). One source of nonindependence, discussed by Strube and Hartmann (1983), arises from multiple hypothesis tests, based upon multiple dependent variable measurements, found within a single study. Three broad options exist for this problem: The meta-analytic reviewer can select and include only the test statistic for the dependent variable measure which is most theoretically relevant to the central issues of the meta-analysis. Alternatively, the meta-analytic reviewer can differentially weight and include the test statistics for all relevant dependent variable measures (assigning heavier weights to the more theoretically relevant measures). Finally, the meta-analytic reviewer can include all statistics and treat them as if they each represented an independent hypothesis test. This latter solution is practical, but risks certain errors (Rosenthal, in press). Whatever option is pursued, the reviewer must make explicit the guidelines used in selecting or weighting results derived from multiple dependent variable measures found within a single study. A detailed discussion of the problem of nonindependence is available elsewhere (Rosenthal, in press).

Decision 3c. Multiple Studies from a Given Laboratory. Another source of nonindependence discussed by Strube and Hartmann (1983) arises from multiple studies conducted by a given researcher or a given research team (cf., Strube & Garcia, 1981). Similarity of results among such related research efforts may be due in part to comon subject populations, common experimental paradigms, or other attributes peculiar to a given laboratory. In such situations, the reviewer might separate studies conducted by different research teams; separate meta-analyses conducted upon these distinct samples of research might reveal effects of this possible source of nonindependence (see also Rosenthal, in press).

Explicit statement of the choices made on each of the decisions outlined in this chapter will make a given application of meta-analytic techniques more readily interpretable and more clearly understood. Further, critique or evaluation of a meta-analysis that makes these choices explicit will be easier, and can be more constructive than has been the case in the past. In consequence, as suggested by Fiske (1983), those researchers who wish to criticize a meta-analysis because of the choice made on one of these decisions will have only to identify the decision in question, and conduct the alternative meta-analysis using the choice which they feel is more defensible for that set of studies.

3 Modifying the Meta-analysis Programs

The programs listed in Chapter 5 are written in a subset of BASIC which is compatible with most popular computer systems. Subtle differences in syntax are found between various "dialects" of BASIC currently in use. For example, in the listings provided here, the operator for exponentiation is "**". This corresponds to "↑" in PET BASIC and in TRS-80 BASIC; however, this corresponds to "λ" in Integer BASIC and Applesoft BASIC. Recently, BASIC computer language dictionaries have been published (e.g., Noonan, 1982), which facilitate translation between versions of BASIC.

The major difference between computer systems which will impact upon the use of these programs is the number of characters per line *(cpl)*. The number of *cpl* for some popular microcomputer systems are: Radio Shack = 65 *cpl;* Apple = 40 *cpl;* Commodore = 40 *cpl;* IBM = 40 *cpl*. The programs listed in Chapter 5 are listed in 72 *cpl* format. When entering these programs into a system with a lower *cpl* format, this will only cause the video display of the program listing to look different. For example, line number 00690 of Program A is listed as:

```
00690 P2 = (.0000380036*(ZM*4)+.0000488906*(ZM***5 )+.000005383*(ZM**6))
```

When this program line is entered into a computer system with 72 *cpl,* the line will be displayed exactly as illustrated above. However, when this program line is entered into an Apple microcomputer (40 *cpl*) the line will be displayed as:

```
00690 P2 = (.0000380036*(ZM**4)+.000048890
6*(ZM**5)+.000005383*(ZM**6))
```

This difference in the videodisplay of entered programs lines (which results from differing *cpl* formats) has no effect on the operation or the statistical output of the programs.

However, a practical problem does arise from these differing *cpl* formats. Although the actual statistical output of these programs is not affected, the *display* of the statistical output is affected. For example, the final display for Program D run on a Radio Shack microcomputer (64 *cpl*) would look like this:

```
 * * * META-ANALYTIC STATISTICS
FISHER Z FOR COMBINATION:                          Z = .329695
MEAN CORRELATION COEFFICIENT:                      R = .318247
PROPORTION OF VARIANCE ACCOUNTED FOR:            R[2 = .101281
BINOMIAL EFFECT SIZE DISPLAY:  FROM  .340877  TO  .659123

MEAN STANDARDIZED DIFFERENCE BETWEEN THE MEANS:D = .671401
```

However, the final display for the same run of Program D on an Apple microcomputer (40 *cpl*) would look like this:

```
    * * * META-ANALYTIC STATISTICS

FISHER Z FOR COMBINATION:
        Z = .329695
MEAN CORRELATION COEFFICIENT:
        R = .318247
PROPORTION OF VARIANCE ACCOUNTED FOR:
      R[2 = .101281
BINOMIAL EFFECT SIZE DISPLAY:  FROM  .
340877  TO  .659123

MEAN STANDARDIZED DIFFERENCE BETWEEN THE
  MEANS:D= .671401
```

Note that the output is identical, but the 40 *cpl* display of the output (and, therefore, its readability) is affected by the reduction in *cpl* capability.

The user may choose to modify the display of the output by substituting (where necessary, in those program lines containing PRINT statements) abbreviations for the labels of the generated meta-analytic statistics. For example, the same final display could be abbreviated as follows to fit within 40 *cpl* format:

```
    * * * META-ANALYTIC STATISTICS
FISHER Z FOR COMBINATION = .329695
MEAN R = .318247
R[2 = .101281
BESD:  .340877  TO  .659123

MEAN D = .671401
```

However, these types of modifications are not necessary, and are suggested only as a means of making the output display more readable.

The user might wish to modify the programs so that output is displayed on a line printer rather than on a videomonitor. Depending on the computer system, all PRINT statements could therefore be changed to LPRINT statements, or to PRINT # statements (where # represents the character number associated with the user's line printer).

Finally, the user might wish to reduce the random access memory requirements of these programs by eliminating blank spaces not required by the user's computer system. For example, in these programs the command "GOTO" is always preceded and followed by a blank space. Some computer systems will require the blank spaces; however, some computer systems will *not* require the blank spaces, and their elimination will save considerable memory capacity.

In all cases, users are advised to consult the operation manual for their own computer system.

4 Input and Output: How to Enter Data

This chapter describes the data which the meta-analysis programs prompt the user to enter, and the output produced by each of the programs. For those programs dealing with significance levels, the user is prompted to enter for each study the type of statistic *(t, F, χ^2, r,* or *p),* the numerical value of the statistic, and the direction of effect (+ if in the expected direction, — if in the unexpected direction). The user may also be prompted to enter the associated *df,* as well as a weight for each study (cf. descriptions of Programs A, C, E, H and J given earlier). For those programs dealing with effect sizes, the user is prompted to enter for each study the type of statistic *(t, F, χ^2, r* or *p),* the numerical value of the statistic, the associated *df* or N, and the direction of effect (+ if in the expected direction, − if in the unexpected direction). The user may also be prompted to enter a weight for each study (cf., the descriptions of Programs B, D, F, I and K above). Note that F can be employed only when $df = 1$ for the numerator; analogously, χ^2 can be employed only when $df = 1$.

After the input of each study's data, the programs will output and display relevant retrieved information (e.g., Z associated with input study data; or, effect size indices derived from input study data). Once the user has entered all of the studies' data, the programs will output and display the associated meta-analytic statistics.

In the examples that follow, reference will be made to the data illustrated in Table 4.1. This table conveys the results of the only five studies that have examined the effect of X on Y (substitute for X and Y any variables of interest to the reader). This hypothetical set of studies has been constructed to illustrate the application of the meta-analytic procedures described pre-

TABLE 4.1
Set of five studies examining the effects of X on Y

Study (Date)	Duration of X used in the study	Study statistic	N	(df)	Significance: one-tailed p value	Significance: Z^a	Effect Size: r	Effect Size: d	Effect Size: Fisher Z
Smith (1980)	8	$X^2 =$ 6.635.	100	(1)	.005	2.57585	.257585	.533161	.26352
Jones (1980)	10	$t =$ 3.46	60	(58)	.0005	3.28476	.413633	.908641	.439986
Johnson (1981)	4	$F =$ 4.18.	31	(1,29)	.025	1.95897	.354936	.75931	.37108
Williams (1982)	6	$r =$ .5368	22	(20)	.005	2.57448	.5368	1.27248	.599649
Brown (1983)	2	Mann-Whitney $U = 25$	n1 = 7, n2 = 8, $N = 15$	(−)	.389	.281499	.0726826	.145751	.0728108

Note: [a]When the Z is positive the result is in the predicted direction. When the Z is negative the result is in the unpredicted direction.

viously. For the sake of illustration, the meta-analytic techniques involving two studies will be applied to the results of the first two studies listed in Table 4.1. When a program prompts the user to enter a contrast weight, the illustrations in the following will employ linear polynomial coefficients derived for "Duration of *X* used in the study" from Program G. When a program prompts a user to enter a weight other than a contrast weight, the illustrations will employ the sample size *(N)* associated with each study. All of the meta-analytic statistics which are output and displayed by the computer programs below will also be computed by hand. Note that computer systems typically "round down" (e.g., ".4999999" would be rounded to six decimal places as ".499999"); this accounts for the slight discrepancies between the computer output and the calculation by hand in the last decimal place for some of the illustrations given.

Program A. Comparison of Significance Levels (Two Studies). The program prompts the user to enter the type of statistic *(t, F,* χ^2, *r* or *p),* the numerical value of the statistic, the associated *df* for some of these statistics, and the direction of effect for each study. The program displays the *Z* associated with the data entered for each study. For example:

```
META-ANALYSIS COMPARING SIGNIFICANCE LEVELS
(TWO STUDIES)

ENTER TYPE OF STATISTIC FOR DATA SET # 1
(T = T-TEST; F = F-TEST; X = CHI-SQUARE; R =
CORRELATION COEF; P = PROBABILITY VALUE)
? X
ENTER CHI-SQUARE FOR DATE SET # 1
? 6.635
ENTER DIRECTION OF EFFECT ('+' IF IN EXPECTED
DIRECTION, '−' IF IN OPPOSITE DIRECTION)
? +
Z = 2.57585
ENTER TYPE OF STATISTIC FOR DATA SET # 2
(T = T-TEST; F = F-TEST; X = CHI-SQUARE; R =
CORRELATION COEF; P = PROBABILITY VALUE)
? T
ENTER T-TEST FOR DATA SET # 2
? 3.46
ENTER DF ASSOCIATED WITH THIS T-TEST
? 58
ENTER DIRECTION OF EFFECT ('+' IF IN EXPECTED
```

```
DIRECTION, '–' IF IN OPPOSITE DIRECTION)
? +
Z = 3.28476
```

The program then displays the computed Z for comparison and the associated one-tailed p level, indicating the likelihood that this difference between the significance levels of the two included studies, or a difference more extreme, could have occurred if the null hypothesis were true. For example:

```
* * * META-ANALYTIC STATISTICS
Z FOR COMPARISON = .501278
ASSOCIATED ONE-TAILED P = .308088
```

Performing the computation of the Z for comparison by hand:

$$Z = \frac{Z1 - Z2}{\sqrt{2}} = \frac{2.57585 - 3.28476}{\sqrt{2}} = .501276$$

Thus, the significance levels generated by Smith (1980) and Jones (1980) are not significantly different.

Program B. Comparison of Effect Sizes (Two Studies). The program prompts the user to enter the type of statistic (t, F, χ^2, r or p), the numerical value of the statistic, the associated *df* or N, and the direction of effect for each study. The program displays the effect size indices derived from the entered data for each study. For example:

```
META-ANALYSIS COMPARING EFFECT SIZES
(TWO STUDIES)

DATA SET # 1
ENTER TYPE OF STATISTIC FOR DATA SET # 1
(T = T-TEST; F = F-TEST; X = CHI-SQUARE; R =
CORRELATION COEF; P = PROBABILITY VALUE)
? X
ENTER CHI-SQUARE FOR DATA SET # 1
? 6.635
ENTER N ASSOCIATED WITH THIS CHI-SQUARE
? 100
ENTER DIRECTION OF EFFECT ('+' IF IN EXPECTED
DIRECTION, '–' IF IN OPPOSITE DIRECTION)
? +
```

```
EFFECT SIZE STATISTICS FOR DATA SET # 1
CORRELATION COEFFICIENT:
        R= .257585
PROPORTION OF VARIANCE ACCOUNTED FOR:
      R[2= .06635
BINOMIAL EFFECT SIZE DISPLAY:  FROM  .371208
TO  .628793

STANDARDIZED DIFFERENCE BETWEEN MEANS:
        D= .533161

ENTER TO CONTINUE?

DATA SET # 2
ENTER TYPE OF STATISTIC FOR DATA SET # 2
(T = T-TEST; F = F-TEST; X = CHI-SQUARE; R =
CORRELATION COEF; P = PROBABILITY VALUE)
? T
ENTER T-TEST FOR DATA SET # 2
? 3.46
ENTER DF ASSOCIATED WITH THIS T-TEST
? 58
ENTER DIRECTION OF EFFECT ('+' IF IN EXPECTED
DIRECTION, '-' IF IN OPPOSITE DIRECTION)
? +

EFFECT SIZE STATISTICS FOR DATA SET # 2
CORRELATION COEFFICIENT:
        R= .413633
PROPORTION OF VARIANCE ACCOUNTED FOR:
      R[2= .171092
BINOMIAL EFFECT SIZE DISPLAY:  FROM  .293183
TO  .706817

STANDARDIZED DIFFERENCE BETWEEN MEANS;
        D= .908641

ENTER TO CONTINUE?
```

The program then displays the computed Z for comparision and the associated p level, indicating the likelihood that this difference between the effect sizes of the two included studies, or one more extreme, could have occurred if the null hypothesis were true. For example:

```
* * * META-ANALYTIC STATISTICS
Z FOR COMPARISON = 1.05737
ASSOCIATED ONE-TAILED P = .145172
```

Performing the computation of the Z for comparison by hand:

$$Z = \frac{Z1 - Z2}{\sqrt{\frac{1}{N1-3} + \frac{1}{N2-3}}} = \frac{.26352 - .439986}{\sqrt{\frac{1}{(100-3)} + \frac{1}{(60-3)}}} = 1.05736$$

Thus, the effect sizes generated by Smith (1980) and Jones (1980) are not significantly different.

Program C. Combination of Significance Levels (Two Studies). The program prompts the user to enter the type of statistic, the numerical value of the statistic, the associated *df* for some of these statistics, the direction of effect, and a weight for each study. The program displays the Z associated with the data entered for each study. For example:

```
META-ANALYSIS COMBINING SIGNIFICANCE LEVELS
                (TWO STUDIES)

ENTER TYPE OF STATISTIC FOR DATA SET # 1
(T = T-TEST; F = F-TEST; X = CHI-SQUARE; R =
CORRELATION COEF; P = PROBABILITY VALUE)
? X
ENTER CHI-SQUARE FOR DATA SET # 1
? 6.635
ENTER DIRECTION OF EFFECT ('+' IF IN EXPECTED
DIRECTION, '-' IF IN OPPOSITE DIRECTION)
? +

ENTER WEIGHT? 100.

Z = 2.57585
ENTER TYPE OF STATISTIC FOR DATA SET # 2
(T = T-TEST; F = F-TEST; X = CHI-SQUARE; R =
CORRELATION COEF; P = PROBABILITY VALUE)
? T
ENTER T-TEST FOR DATA SET # 2
? 3.46
ENTER DF ASSOCIATED WITH THIS T-TEST
? 58
ENTER DIRECTION OF EFFECT ('+' IF IN EXPECTED
```

```
DIRECTION, '−' IF IN OPPOSITE DIRECTION)
? +

ENTER WEIGHT? 60

Z = 3.28476
```

Note that the weights entered here are the sample sizes associated with each study. The program then displays the computed Z for combination and the associated p level, indicating the likelihood that the combined results of the two included studies, or one more extreme, could have been obtained if the null hypothesis were true. For example:

```
* * * META-ANALYTIC STATISTICS
Z FOR COMBINATION = 3.89877
ASSOCIATED ONE-TAILED P = 4.8467E-05
```

Performing the computation of the Z for combination by hand:

$$Z = \frac{w1Z1 + w2Z2}{\sqrt{w1^2 + w2^2}} = \frac{100\,(2.57585) + 60\,(3.28476)}{\sqrt{(100)^2 + (60)^2}} = 3.89877$$

Thus the combined results of Smith (1980) and Jones (1980) are unlikely to have occurred by chance.

Program D. Combination of Effect Sizes (Two Studies). The program prompts the user to enter the type of statistic, the numerical value of the statistic, the associated *df* or N, the direction of effect, and a weight for each study. The program displays the effect size indices derived from the entered data for each study. For example:

```
META-ANALYSIS COMBINING EFFECT SIZES
(TWO STUDIES)

DATA SET # 1
ENTER TYPE OF STATISTIC FOR DATA SET # 1
(T = T-TEST; F = F-TEST; X = CHI-SQUARE; R =
CORRELATION COEF; P = PROBABILITY VALUE)
? X
ENTER CHI-SQUARE FOR DATA SET # 1
? 6.635
ENTER N ASSOCIATED WITH THIS CHI-SQUARE
? 100
ENTER WEIGHT
```

```
? 100
ENTER DIRECTION OF EFFECT ('+' IF IN EXPECTED
DIRECTION, '-' IF IN OPPOSITE DIRECTION)
? +
EFFECT SIZE STATISTICS FOR DATA SET # 1
CORRELATION COEFFICIENT:
          R= .257585
PROPORTION OF VARIANCE ACCOUNTED FOR:
        R[2= .06635
BINOMIAL EFFECT SIZE DISPLAY:   FROM   .371208
TO   .628793

STANDARDIZED DIFFERENCE BETWEEN MEANS:
          D= .533161

ENTER TO CONTINUE?

DATA SET # 2
ENTER TYPE OF STATISTIC FOR DATA SET # 2
(T = T-TEST; F = F-TEST; X = CHI-SQUARE; R =
CORRELATION COEF; P = PROBABILITY VALUE)
? T
ENTER T-TEST FOR DATA SET # 2
? 3.46
ENTER DF ASSOCIATED WITH THIS T-TEST
? 58
ENTER WEIGHT
? 60
ENTER DIRECTION OF EFFECT ('+' IF IN EXPECTED
DIRECTION, '-' IF IN OPPOSITE DIRECTION)
? +

EFFECT SIZE STATISTICS FOR DATA SET # 2
CORRELATION COEFFICIENT:
          R= .413633
PROPORTION OF VARIANCE ACCOUNTED FOR:
        R[2= .171092
BINOMIAL EFFECT SIZE DISPLAY:   FROM   .293183
TO   .706817

STANDARDIZED DIFFERENCE BETWEEN MEANS:
          D= .908641

ENTER TO CONTINUE?
```

Note that the weights entered here are the sample sizes associated with each study. The program then displays the mean indices of effect size, indicating the combined effect size for the two included studies. For example:

```
  *** META-ANALYTIC STATISTICS
FISHER Z FOR COMBINATION:
        Z= .329695
MEAN CORRELATION COEFFICIENT:
        R= .318247
PROPORTION OF VARIANCE ACCOUNTED FOR:
        R[2= .101281
BINOMIAL EFFECT SIZE DISPLAY:  FROM  .340877
TO  .659123

MEAN STANDARDIZED DIFFERENCE BETWEEN THE
MEANS:D = .671401
```

Performing the computation of the *Z* for combination by hand:

$$Z = \frac{w1Z1 + w2Z2}{w1 + w2} = \frac{100\,(.26352) + 60\,(.439986)}{100 + 60} = .329695$$

Thus the combined effect size obtained from Smith (1980) and Jones (1980) appears to be moderate in magnitude.

Program E. Comparison of Significance Levels (Diffuse Test) (Three or more Studies). The program prompts the user to enter the number of studies to be included, the type of statistic, the numerical value of the statistic, the associated *df* for some of these statistics, and the direction of effect for each study. The program displays the *Z* associated with the data entered for each study. For example:

```
        META-ANALYSIS COMPARING SIGNIFICANCE
                LEVELS (DIFFUSE TESTS)
               (THREE OR MORE STUDIES)

ENTER THE NUMBER OF STUDIES TO BE INCLUDED?
5
ENTER TYPE OF STATISTIC FOR DATA SET # 1
(T = T-TEST; F = F-TEST; X = CHI-SQUARE; R =
CORRELATION COEF; P = PROBABILITY VALUE)
? X
ENTER CHI-SQUARE FOR DATA SET # 1
```

```
? 6.635
ENTER DIRECTION OF EFFECT ('+' IF IN EXPECTED
DIRECTION, '-' IF IN OPPOSITE DIRECTION)
? +

Z = 2.57585
ENTER TYPE OF STATISTIC FOR DATA SET # 2
(T = T-TEST; F = F-TEST; X = CHI-SQUARE; R =
CORRELATION COEF; P = PROBABILITY VALUE)
? T
ENTER T-TEST FOR DATA SET # 2
? 3.46
ENTER DF ASSOCIATED WITH THIS T-TEST
? 58
ENTER DIRECTION OF EFFECT ('+' IF IN EXPECTED
DIRECTION, '-' IF IN OPPOSITE DIRECTION)
? +

Z = 3.28476
ENTER TYPE OF STATISTIC FOR DATA SET # 3
(T = T-TEST; F = F-TEST; X = CHI-SQUARE; R =
CORRELATION COEF; P = PROBABILITY VALUE)
? F
ENTER F-TEST FOR DATA SET # 3
? 4.18
ENTER DF(ERROR) ASSOCIATED WITH THIS F-TEST
? 29
ENTER DIRECTION OF EFFECT ('+' IF IN EXPECTED
DIRECTION, '-' IF IN OPPOSITE DIRECTION)
? +

Z = 1.95897
ENTER TYPE OF STATISTIC FOR DATA SET # 4
(T = T-TEST; F = F-TEST; X = CHI-SQUARE; R =
CORRELATION COEF; P = PROBABILITY VALUE)
? R
ENTER CORRELATION COEFFICIENT FOR DATA SET #
4
? .5368
ENTER DF ASSOCIATED WITH THIS CORRELATION
```

```
? 20
ENTER DIRECTION OF EFFECT ('+' IF IN EXPECTED
DIRECTION, '-' IF IN OPPOSITE DIRECTION)
? +

Z = 2.57448
ENTER TYPE OF STATISTIC FOR DATA SET # 5
(T = T-TEST; F = F-TEST; X = CHI-SQUARE; R =
CORRELATION COEF; P = PROBABILITY VALUE)
? P
ENTER ONE-TAILED P FOR DATA SET # 5
? .389
Z = .281499
```

The program then displays the computed χ^2 for comparison with k-1 degrees of freedom and the associated p level. For example:

```
* * * META-ANALYTIC STATISTICS
CHI-SQUARE FOR COMPARISON = 5.1759
DF = 4
ASSOCIATED ONE-TAILED P = .26972
```

Performing the computation of the χ^2 for comparison by hand first requires the computation of the mean Z:

$$\bar{Z} = \frac{\Sigma Zj}{k} = \frac{2.57585 + 3.28476 + 1.95897 + 2.57448 + .281499}{5} = 2.13511$$

Then:

$$\chi^2_{(k-1)} = \Sigma(Zj-\bar{Z})^2 = (2.57585-2.13511)^2 + (3.28476-2.13511)^2 + (1.95897-2.13511)^2 + (2.57448-2.13511)^2 + (.281499-2.13511)^2 = 5.17589$$

Thus the significance levels of the five studies listed in Table 4.1 do not appear to be significantly heterogeneous.

Program F. Comparison of Effect Sizes (Diffuse Test) (Three or more Studies). The program prompts the user to enter the number of studies to be included, the type of statistic, the numerical value of the statistic, the associated *df* or N, and the direction of effect for each study. The program displays the effect size indices derived from the entered data for each study. For example:

META-ANALYSIS COMPARING EFFECT SIZES
(DIFFUSE TESTS)
(THREE OR MORE STUDIES)

ENTER THE NUMBER OF STUDIES TO BE INCLUDED?
5
DATA SET # 1
ENTER TYPE OF TEST STATISTIC FOR DATA SET # 1
(T = T-TEST; F = F-TEST; X = CHI-SQUARE; R = CORRELATION COEF; P = PROBABILITY VALUE)
? X
ENTER CHI-SQUARE STATISTIC FOR DATA SET # 1
? 6.635
ENTER N ASSOCIATED WITH THIS CHI-SQUARE
? 100
ENTER DIRECTION OF EFFECT ('+' IF IN EXPECTED DIRECTION, '−' IF IN OPPOSITE DIRECTION)
? +

EFFECT SIZE STATISTICS FOR DATA SET # 1
CORRELATION COEFFICIENT:
R= .257585
PROPORTION OF VARIANCE ACCOUNTED FOR:
R[2= .06635
BINOMIAL EFFECT SIZE DISPLAY: FROM .371208 TO .628793

STANDARDIZED DIFFERENCE BETWEEN MEANS:
D= .533161

ENTER TO CONTINUE?

DATA SET # 2
ENTER TYPE OF TEST STATISTIC FOR DATA SET # 2
(T = T-TEST; F = F-TEST; X = CHI-SQUARE; R = CORRELATION COEF; P = PROBABILITY VALUE)
? T
ENTER T-TEST STATISTIC FOR DATA SET # 2
? 3.46
ENTER DF ASSOCIATED WITH THIS T-TEST
? 58
ENTER DIRECTION OF EFFECT ('+' IF IN EXPECTED DIRECTION, '−' IF IN OPPOSITE DIRECTION)
? +

EFFECT SIZE STATISTICS FOR DATA SET # 2
CORRELATION COEFFICIENT:
R= .413633
PROPORTION OF VARIANCE ACCOUNTED FOR:
R[2= .171092
BINOMIAL EFFECT SIZE DISPLAY: FROM .293183
TO .706817

STANDARDIZED DIFFERENCE BETWEEN MEANS:
D= .908641

ENTER TO CONTINUE?

DATA SET # 3
ENTER TYPE OF TEST STATISTIC FOR DATA SET # 3
(T = T-TEST; F = F-TEST; X = CHI-SQUARE; R = CORRELATION COEF; P = PROBABILITY VALUE)
? F
ENTER F-TEST STATISTIC FOR DATA SET # 3
? 4.18
ENTER DF(ERROR) ASSOCIATED WITH THIS F-TEST
? 29
ENTER DIRECTION OF EFFECT ('+' IF IN EXPECTED DIRECTION, '−' IF IN OPPOSITE DIRECTION)
? +

EFFECT SIZE STATISTICS FOR DATA SET # 3
CORRELATION COEFFICIENT:
R= .354936
PROPORTION OF VARIANCE ACCOUNTED FOR:
R[2= .125979
BINOMIAL EFFECT SIZE DISPLAY: FROM
.322532 TO .677468

STANDARDIZED DIFFERENCE BETWEEN MEANS:
D= .75931

ENTER TO CONTINUE?

DATA SET # 4
ENTER TYPE OF TEST STATISTIC FOR DATA SET # 4
(T = T-TEST; F = F-TEST; X = CHI-SQUARE; R = CORRELATION COEF; P = PROBABILITY VALUE)
? R
ENTER CORRELATION COEFFICIENT FOR DATA SET #

```
4
? .5368
ENTER DF ASSOCIATED WITH THIS CORRELATION
COEFFICIENT
? 20
ENTER DIRECTION OF EFFECT ('+' IF IN EXPECTED
DIRECTION, '-' IF IN OPPOSITE DIRECTION)
? +

EFFECT SIZE STATISTICS FOR DATA SET # 4
CORRELATION COEFFICIENT:
        R= .5368
PROPORTION OF VARIANCE ACCOUNTED FOR:
      R[2= .288154
BINOMIAL EFFECT SIZE DISPLAY:  FROM
.2316  TO  .7684

STANDARDIZED DIFFERENCE BETWEEN MEANS:
        D= 1.27248

ENTER TO CONTINUE?

DATA SET # 5
ENTER TYPE OF TEST STATISTIC FOR DATA SET # 5
(T = T-TEST; F = F-TEST; X = CHI-SQUARE; R =
CORRELATION COEF; P = PROBABILITY VALUE)
? P
ENTER PROBABILITY VALUE FOR DATA SET # 5
? .389
ENTER N ASSOCIATED WITH THIS PROBABILITY
VALUE
? 15
ENTER DIRECTION OF EFFECT ('+' IF IN EXPECTED
DIRECTION, '-' IF IN OPPOSITE DIRECTION)
? +

EFFECT SIZE STATISTICS FOR DATA SET # 5
CORRELATION COEFFICIENT:
        R= .0726826
PROPORTION OF VARIANCE ACCOUNTED FOR:
      R[2= 5.28277E-03
BINOMIAL EFFECT SIZE DISPLAY:  FROM
.463659  TO  .536341
```

STANDARDIZED DIFFERENCE BETWEEN MEANS:
D= .145751

ENTER TO CONTINUE?

The program then displays the computed χ^2 for comparison with k-1 degrees of freedom and the associated p level. For example:

★ ★ ★ META-ANALYTIC STATISTICS
CHI-SQUARE FOR COMPARISON = 3.29827
DF = 4
ASSOCIATED ONE-TAILED P = .509207

Performing the computation of the χ^2 for comparison by hand first requires the computation of the weighted mean Z:

$$\text{weighted } \bar{Z} = \frac{\Sigma \text{wj}Z\text{j}}{\Sigma \text{wj}} = \frac{100\,(.26352) + 60\,(.439986) + 31(.37108) + 22\,(.599649) + 15\,(.0728108)}{100 + 60 + 31 + 22 + 15} = .34447$$

Then:

$$\chi^2_{(k-1)} = \Sigma(N\text{j} - 3)(Z\text{j} - \bar{Z})^2 = (100-3)\,(.26352 - .34447)^2 + (60-3)\,(.439986 - .34447)^2 + (31-3)\,(.37108 - .34447)^2 + (22-3)\,(.599649 - .34447)^2 + (15-3)\,(.0728108 - .34447)^2 = 3.29827$$

Thus the effect sizes of the five studies listed in Table 4.1 do not appear to be significantly heterogeneous.

Program G. Linear and Quadratic Orthogonal Contrast Weights. This program computes linear and quadratic orthogonal coefficients that can be used as contrast weights to test linear and quadratic trends in study outcomes using Programs H and I. These contrast weights are derived from study attributes (e.g., duration of independent variable manipulation, date of publication, etc.), and are "tailored" to the set of studies used to derive them. These contrast weights do not require equal spacing between levels of the study attribute, nor do they require equal numbers of exemplars at the various levels of the study attribute. The user is prompted to enter the number of levels of the study attribute, the numeric value of each level of the study attribute, and the number of exemplars at each level of the study attribute. Linear and quadratic contrast weights are then computed (using the procedure described in Kirk, 1982, Appendix C, pp. 773–776) and displayed.

For example, using the attribute of "Duration of *X* used in the study," there are five levels of the attribute (i.e., 10, 8, 6, 4 and 2), with one exemplar at each level. Note that the levels are entered in rank order (levels of the attribute may be entered in ascending or descending rank order, but they must be entered in order):

```
COMPUTATION OF ORTHOGONAL COEFFICIENTS
(LINEAR AND QUADRATIC)

ENTER NUMBER OF LEVELS OF ATTRIBUTE?  5
ENTER NUMERICAL VALUE OF ATTRIBUTE LEVEL  1
? 10
ENTER NUMBER OF EXEMPLARS OF ATTRIBUTE
LEVEL  1
? 1

ENTER NUMERICAL VALUE OF ATTRIBUTE LEVEL  2
? 8
ENTER NUMBER OF EXEMPLARS OF ATTRIBUTE
LEVEL  2
? 1

ENTER NUMERICAL VALUE OF ATTRIBUTE LEVEL  3
? 6
ENTER NUMBER OF EXEMPLARS OF ATTRIBUTE
LEVEL  3
? 1

ENTER NUMERICAL VALUE OF ATTRIBUTE LEVEL  4
? 4
ENTER NUMBER OF EXEMPLARS OF ATTRIBUTE
LEVEL  4
? 1

ENTER NUMERICAL VALUE OF ATTRIBUTE LEVEL  5
? 2
ENTER NUMBER OF EXEMPLARS OF ATTRIBUTE
LEVEL  5
? 1
```

The final display instructs the user that the derived coefficients may be simplified by some common denominator (e.g., if all coefficients are even, then the coefficients can be simplified by dividing them by "2"). The displayed coefficients for "Duration of *X* used in the study" listed in Table 4.1 are:

```
ORTHOGONAL COEFFICIENTS
USER MAY SIMPLIFY COEFFICIENTS BY SOME COM-
MON DENOMINATOR (E.G., DIVIDE EVEN COEFFI-
CIENTS BY 2)
LINEAR COEFFICIENT FOR LEVEL  1     =   20
QUADRATIC COEFFICIENT FOR LEVEL  1  =   8

ENTER TO CONTINUE?
LINEAR COEFFICIENT FOR LEVEL  2     =   10
QUADRATIC COEFFICIENT FOR LEVEL  2  = -4.00001

ENTER TO CONTINUE?
LINEAR COEFFICIENT FOR LEVEL  3     =   0
QUADRATIC COEFFICIENT FOR LEVEL  3  =  -8

ENTER TO CONTINUE?
LINEAR COEFFICIENT FOR LEVEL  4     =  -10
QUADRATIC COEFFICIENT FOR LEVEL  4  =  -4

ENTER TO CONTINUE?
LINEAR COEFFICIENT FOR LEVEL  5     =  -20
QUADRATIC COEFFICIENT FOR LEVEL  5  =   8
```

The linear coefficients provided above (20, 10, 0, −10, −20) might be simplified by dividing by "10" (yielding 2, 1, 0, −1, −2); both sets of coefficients will produce the same results when used as contrast weights in Programs H and I (in the illustrations of Programs H and I which follow, the simplified weights [2, 1, 0, −1, −2] will be used).

Note that if a data set involves unequal spacing between levels of the study attribute, or if a data set involves unequal numbers of exemplars at each level of the study attribute, the derived coefficients may be extended decimal numbers. For example, using the study attribute of "Date of publication," there are four levels of the attribute (1980, 1981, 1982, and 1983), with two exemplars at the first level and one exemplar at each of the remaining three levels. For this attribute, the numeric values are transformed to the smallest integer, interval scale, in order to reduce the computed contrast weights to a manageable metric. Thus; 1980 = 1; 1981 = 2; 1982 = 3; 1983 = 4. The quadratic coefficients derived for this study attribute are: 1980 (two exemplars) yields .588235; 1981 (one exemplar) yields −1.23529; 1982 (one exemplar) yields −1.05882; and, 1983 (one exemplar) yields 1.11765.

Program H. Comparison of Significance Levels (Focused Test) (Three or more Studies). The program prompts the user to enter the number of studies to be included, the type of statistic, the numerical value of the statistic, the associated *df* for some of these statistics, the direction of effect, and a contrast weight for each study. The program displays the *Z* associated with the data entered for each study. For example:

```
META-ANALYSIS COMPARING SIGNIFICANCE LEVELS
              (FOCUSED TESTS)
          (THREE OR MORE STUDIES)

ENTER THE NUMBER OF STUDIES TO BE INCLUDED?
5
ENTER TYPE OF STATISTIC FOR DATA SET # 1
(T = T-TEST; F = F-TEST; X = CHI-SQUARE; R =
CORRELATION COEF; P = PROBABILITY VALUE)
? X
ENTER CHI-SQUARE FOR DATA SET # 1
? 6.635
ENTER DIRECTION OF EFFECT ('+' IF IN EXPECTED
DIRECTION, '-' IF IN OPPOSITE DIRECTION)
? +

ENTER CONTRAST WEIGHT? 1
Z = 2.57585
ENTER TYPE OF STATISTIC FOR DATA SET # 2
(T = T-TEST; F = F-TEST; X = CHI-SQUARE; R =
CORRELATION COEF; P = PROBABILITY VALUE)
? T
ENTER T-TEST FOR DATA SET # 2
? 3.46
ENTER DF ASSOCIATED WITH THIS T-TEST
? 58
ENTER DIRECTION OF EFFECT ('+' IF IN EXPECTED
DIRECTION, '-' IF IN OPPOSITE DIRECTION)
? +

ENTER CONTRAST WEIGHT? 2
Z = 3.28476
ENTER TYPE OF STATISTIC FOR DATA SET # 3
(T = T-TEST; F = F-TEST; X = CHI-SQUARE; R =
CORRELATION COEF; P = PROBABILITY VALUE)
```

```
? F
ENTER F-TEST FOR DATA SET # 3
? 4.18
ENTER DF (ERROR) ASSOCIATED WITH THIS F-TEST
? 29
ENTER DIRECTION OF EFFECT ('+' IF IN EXPECTED
DIRECTION, '-' IF IN OPPOSITE DIRECTION)
? +

ENTER CONTRAST WEIGHT? -1
Z = 1.95897
ENTER TYPE OF STATISTIC FOR DATA SET # 4
(T = T-TEST; F = F-TEST; X = CHI-SQUARE; R =
CORRELATION COEF; P = PROBABILITY VALUE)
? R
ENTER CORRELATION COEFFICIENT FOR DATA SET #
4
? .5368
ENTER DF ASSOCIATED WITH THIS CORRELATION
? 20
ENTER DIRECTION OF EFFECT ('+' IF IN EXPECTED
DIRECTION, '-' IF IN OPPOSITE DIRECTION)
? +

ENTER CONTRAST WEIGHT? 0
Z = 2.57448
ENTER TYPE OF STATISTIC FOR DATA SET # 5
(T = T-TEST; F = F-TEST; X = CHI-SQUARE; R =
CORRELATION COEF; P = PROBABILITY VALUE)
? P
ENTER ONE-TAILED P FOR DATA SET # 5
? .389

ENTER CONTRAST WEIGHT? -2
Z = .281499
```

Note that the weights entered here are the simplified contrast weights derived for "Duration of *X* used in the study" from Program G above. The program then displays the *Z* for comparison and the associated *p* level, indicating the likelihood that the significance levels of the included studies are significantly predictable in the manner specified by the contrast weights. For example:

```
* * * META-ANALYTIC STATISTICS
Z FOR COMPARISON = 2.09451
ASSOCIATED ONE-TAILED P = .0182149
```

Performing the computation of the Z for comparison by hand:

$$Z = \frac{\Sigma \lambda j Zj}{\sqrt{\Sigma \lambda j^2}} = \frac{1\,(2.57585) + 2\,(3.28476) - 1\,(1.95897) + 0\,(2.57448) - 2\,(.281499)}{\sqrt{(1)^2 + (2)^2 + (-1)^2 + (0)^2 + (-2)^2}}$$

$$= 2.09451$$

Thus, the significance levels of the five studies listed in Table 4.1 do appear to decrease as the duration of X used in the study increases (i.e., longer durations of X produce more significant results). Note that this focused test comparison reveals a significant trend in significance levels, whereas the results of the diffuse test comparison of these same data (Program E, described above) revealed no significant overall heterogeneity. This highlights the power of the focused test comparison to illuminate and substantiate underlying differences which might otherwise go unheeded.

Program I. Comparison of Effect Sizes (Focused Test) (Three or more Studies). The program prompts the user to enter the number of studies to be included, the type of statistic, the numerical value of the statistic, the associated *df* or *N,* the direction of effect, and a contrast weight for each study. For example:

```
          META-ANALYSIS COMPARING EFFECT SIZES
                   (FOCUSED TESTS)
              (THREE OR MORE STUDIES)

ENTER THE NUMBER OF STUDIES TO BE INCLUDED?
5
DATA SET # 1
ENTER TYPE OF STATISTIC FOR DATA SET # 1
(T = T-TEST; F = F-TEST; X = CHI-SQUARE; R =
CORRELATION COEF; P = PROBABILITY VALUE)
? X
ENTER CHI-SQUARE FOR DATA SET # 1
? 6.635
ENTER N ASSOCIATED WITH THIS CHI-SQUARE
? 100
ENTER DIRECTION OF EFFECT ('+' IF IN EXPECTED
```

```
DIRECTION, '-' IF IN OPPOSITE DIRECTION)
? +

ENTER CONTRAST WEIGHT? 1

EFFECT SIZE STATISTICS FOR DATA SET # 1
CORRELATION COEFFICIENT:
        R= .257585
PROPORTION OF VARIANCE ACCOUNTED FOR:
      R[2= .06635
BINOMIAL EFFECT SIZE DISPLAY:   FROM
.371208   TO   .628793

STANDARDIZED DIFFERENCE BETWEEN MEANS:
        D= .533161

ENTER TO CONTINUE?

DATA SET # 2
ENTER TYPE OF STATISTIC FOR DATA SET # 2
(T = T-TEST; F = F-TEST; X = CHI-SQUARE; R =
CORRELATION COEF; P = PROBABILITY VALUE)
? T
ENTER T-TEST FOR DATA SET # 2
? 3.46
ENTER DF ASSOCIATED WITH THIS T-TEST
? 58
ENTER DIRECTION OF EFFECT ('+' IF IN EXPECTED
DIRECTION, '-' IF IN OPPOSITE DIRECTION)
? +

ENTER CONTRAST WEIGHT? 2

EFFECT SIZE STATISTICS FOR DATA SET # 2
CORRELATION COEFFICIENT:
        R= .413633
PROPORTION OF VARIANCE ACCOUNTED FOR:
      R[2= .171092
BINOMIAL EFFECT SIZE DISPLAY:   FROM
.293183   TO   .706817

STANDARDIZED DIFFERENCE BETWEEN MEANS:
        D= .908641
```

```
ENTER TO CONTINUE?

DATA SET # 3
ENTER TYPE OF STATISTIC FOR DATA SET # 3
(T = T-TEST; F = F-TEST; X = CHI-SQUARE; R =
CORRELATION COEF; P = PROBABILITY VALUE)
? F
ENTER F-TEST FOR DATA SET # 3
? 4.18
ENTER DF(ERROR) ASSOCIATED WITH THIS F-TEST
? 29
ENTER DIRECTION OF EFFECT ('+' IF IN EXPECTED
DIRECTION, '-' IF IN OPPOSITE DIRECTION)
? +

ENTER CONTRAST WEIGHT? -1

EFFECT SIZE STATISTICS FOR DATA SET # 3
CORRELATION COEFFICIENT:
        R= .354936
PROPORTION OF VARIANCE ACCOUNTED FOR:
      R[2= .125979
BINOMIAL EFFECT SIZE DISPLAY:   FROM
.322532   TO   .677468

STANDARDIZED DIFFERENCE BETWEEN MEANS:
        D= .75931

ENTER TO CONTINUE?

DATA SET # 4
ENTER TYPE OF STATISTIC FOR DATA SET # 4
(T = T-TEST; F = F-TEST; X = CHI-SQUARE; R =
CORRELATION COEF; P = PROBABILITY VALUE)
? R
ENTER CORRELATION COEFFICIENT FOR DATA SET #
4
? .5368
ENTER DF ASSOCIATED WITH THIS CORRELATION
COEFFICIENT
? 20
ENTER DIRECTION OF EFFECT ('+' IF IN EXPECTED
DIRECTION, '-' IF IN OPPOSITE DIRECTION)
? +
```

```
ENTER CONTRAST WEIGHT? 0

EFFECT SIZE STATISTICS FOR DATA SET # 4
CORRELATION COEFFICIENT:
        R= .5368
PROPORTION OF VARIANCE ACCOUNTED FOR:
      R[2= .288154
BINOMIAL EFFCT SIZE DISPLAY:   FROM
.2316   TO   .7684

STANDARDIZED DIFFERENCE BETWEEN MEANS:
        D= 1.27248

ENTER TO CONTINUE?

DATA SET # 5
ENTER TYPE OF STATISTIC FOR DATA SET # 5
(T = T-TEST; F = F-TEST; X = CHI-SQUARE; R =
CORRELATION COEF; P = PROBABILITY VALUE)
? P
ENTER PROBABILITY VALUE FOR DATA SET # 5
? .389
ENTER N ASSOCIATED WITH THIS PROBABILITY
VALUE
? 15
ENTER DIRECTION OF EFFECT ('+' IF IN EXPECTED
DIRECTION, '-' IF IN OPPOSITE DIRECTION)
? +

ENTER CONTRAST WEIGHT? -2

EFFECT SIZE STATISTICS FOR DATA SET # 5
CORRELATION COEFFICIENT:
        R= .0726826
PROPORTION OF VARIANCE ACCOUNTED FOR:
      R[2= 5.28277E-03
BINOMIAL EFFECT SIZE DISPLAY:   FROM
.463659   TO   .536341

STANDARDIZED DIFFERENCE BETWEEN MEANS:
        D= .145751

ENTER TO CONTINUE?
```

Note that the weights entered here are the simplified contrast weights derived for "Duration of *X* used in the study" from Program G above. The program then displays the computed *Z* for comparison and the associated *p* level, indicating the likelihood that these effect sizes are significantly predictable in the manner specified by the contrast weights. For example:

```
* * * META-ANALYTIC STATISTICS
Z FOR COMPARISON = .934851
ASSOCIATED ONE-TAILED P = .174932
```

Performing the computation of the *Z* for comparison by hand:

$$Z = \frac{\Sigma\lambda jZj}{\sqrt{\sum\frac{(\lambda j^2)}{(Nj-3)}}} = \frac{1\,(.26352)+2\,(.439986)-1\,(.37108)+0\,(.599649)-2\,(.0728108)}{\sqrt{\frac{(1)^2}{(100-3)}+\frac{(2)^2}{(60-3)}+\frac{(-1)^2}{(31-3)}+\frac{(0)^2}{(22-3)}+\frac{(-2)^2}{(15-3)}}}$$

$$= .934851$$

Thus, there appears to be only a slight, and nonsignificant, tendency for the effect sizes of the five studies listed in Table 4.1 to increase with duration of *X* used in the study. Reconciling the results of this focused test comparison (based on Program I) with the results of the previous focused test comparison (based on Program H), it appears that within this domain of research longer durations of *X* produce more significant results, although they do not necessarily produce effects of greater magnitude. Since significance levels depend not only on effect size but also on sample size, it may often occur that contrasts in significance levels will yield results different from contrasts in effect sizes (Rosenthal & Rosnow, 1984b).

Program J. Combination of Significance Levels (Three or more Studies). The program prompts the user to enter the number of studies to be included, the type of statistic, the numerical value of the statistic, the associated *df* for some of these statistics, the direction of effect, and a weight for each study. The program displays the *Z* associated with the data entered for each study. For example:

```
    META-ANALYSIS COMBINING SIGNIFICANCE
      LEVELS (THREE OR MORE STUDIES)

ENTER THE NUMBER OF STUDIES TO BE INCLUDED?
5
ENTER TYPE OF STATISTIC FOR DATA SET # 1
(T = T-TEST; F = F-TEST; X = CHI-SQUARE; R =
CORRELATION COEF; P = PROBABILITY VALUE)
```

```
? X
ENTER CHI-SQUARE FOR DATA SET # 1
? 6.635
ENTER DIRECTION OF EFFECT ('+' IF IN EXPECTED
DIRECTION, '-' IF IN OPPOSITE DIRECTION)
? +

ENTER WEIGHT? 100
Z = 2.57585
ENTER TYPE OF STATISTIC FOR DATA SET # 2
(T = T-TEST; F = F-TEST; X = CHI-SQUARE; R =
CORRELATION COEF; P = PROBABILITY VALUE)
? T
ENTER T-TEST FOR DATA SET # 2
? 3.46
ENTER DF ASSOCIATED WITH THIS T-TEST
? 58
ENTER DIRECTION OF EFFECT ('+' IF IN EXPECTED
DIRECTION, '-' IF IN OPPOSITE DIRECTION)
? +

ENTER WEIGHT? 60
Z = 3.28476
ENTER TYPE OF STATISTIC FOR DATA SET # 3
(T = T-TEST; F = F-TEST; X = CHI-SQUARE; R =
CORRELATION COEF; P = PROBABILITY VALUE)
? F
ENTER F-TEST FOR DATA SET # 3
? 4.18
ENTER DF(ERROR) ASSOCIATED WITH THIS F-TEST
? 29
ENTER DIRECTION OF EFFECT ('+' IF IN EPXECTED
DIRECTION, '-' IF IN OPPOSITE DIRECTION)
? +

ENTER WEIGHT? 31

Z = 1.95897
ENTER TYPE OF STATISTIC FOR DATA SET # 4
(T = T-TEST; F = F-TEST; X = CHI-SQUARE; R =
CORRELATION COEF; P = PROBABILITY VALUE)
? R
ENTER CORRELATION COEFFICIENT FOR DATA SET #
```

```
4
? .5368
ENTER DF ASSOCIATED WITH THIS CORRELATION
? 20
ENTER DIRECTION OF EFFECT ('+' IF IN EXPECTED
DIRECTION, '-' IF IN OPPOSITE DIRECTION)
? +

ENTER WEIGHT? 22

Z = 2.57448
ENTER TYPE OF STATISTIC FOR DATA SET # 5
(T = T-TEST; F = F-TEST; X = CHI-SQUARE; R =
CORRELATION COEF; P = PROBABILITY VALUE)
? P
ENTER ONE-TAILED P FOR DATA SET # 5
? .389

ENTER WEIGHT? 15
Z = .281499
```

Note that the weights entered here are the sample sizes associated with each study. The program then displays the computed Z for combination and the associated p level, indicating the likelihood that a Z this large or larger could have been obtained if the null hypothesis were true. The program also displays a "fail safe" number for the $p = .05$ level of significance. This estimates the number of unretrieved studies averaging null results that would be required to bring the overall combined probability level to the "just significant" convention of $p = .05$. For example:

```
* * * META-ANALYTIC STATISTICS
Z FOR COMBINATION = 4.66336
ASSOCIATED ONE-TAILED P = 1.59122E-06
FAIL SAFE NUMBER (P = .05) = 37.1163
```

Performing the computation of the Z for combination by hand:

$$Z = \frac{\Sigma wjZj}{\sqrt{\Sigma wj^2}} = \frac{100\,(2.57585) + 60\,(3.28476) + 31\,(1.95897) + 22\,(2.57448) + 15\,(.281499)}{\sqrt{(100)^2 + (60)^2 + (31)^2 + (22)^2 + (15)^2}}$$

$$= 4.66336$$

Performing the computation of the "fail safe" number by hand:

$$N_{(p=.05)} = \left(\frac{\Sigma Zj}{1.645}\right)^2 - k = \left(\frac{2.57585 + 3.28476 + 1.95897 + 2.57448 + .281499}{1.645}\right)^2 - 5$$
$$= 37.1163$$

Thus, the combined results of the five studies listed in Table 4.1 were unlikely to have occurred if the null hypothesis were true; approximately thirty-seven additional studies averaging null results would be necessary to bring the overall combined probability level to $p = .05$.

Program K. Combination of Effect Sizes (Three or more Studies). The program prompts the user to enter the number of studies to be included, the type of statistic, the numerical value of the statistic, the associated *df* or *N*, the direction of effect, and a weight for each study. The program displays the effect size indices derived from the entered data for each study. For example:

```
        META-ANALYSIS COMBINING EFFECT SIZES
              (THREE OR MORE STUDIES)

ENTER THE NUMBER OF STUDIES TO BE INCLUDED?
5
DATA SET # 1
ENTER TYPE OF STATISTIC FOR DATA SET # 1
(T = T-TEST; F = F-TEST; X = CHI-SQUARE; R =
CORRELATION COEF; P = PROBABILITY VALUE)
? X
ENTER CHI-SQUARE FOR DATA SET # 1
? 6.635
ENTER N ASSOCIATED WITH THIS CHI-SQUARE
? 100
ENTER DIRECTION OF EFFECT ('+' IF IN EXPECTED
DIRECTION, '-' IF IN OPPOSITE DIRECTION)
? +

ENTER WEIGHT? 100

EFFECT SIZE STATISTICS FOR DATA SET # 1
CORRELATION COEFFICIENT:
        R= .257585
PROPORTION OF VARIANCE ACCOUNTED FOR:
     R[2= .06635
```

```
BINOMIAL EFFECT SIZE DISPLAY:   FROM
.371208   TO   .628793

STANDARDIZED DIFFERENCE BETWEEN MEANS:
          D= .533161

ENTER TO CONTINUE?

DATA SET # 2
ENTER TYPE OF STATISTIC FOR DATA SET # 2
(T = T-TEST; F = F-TEST; X = CHI-SQUARE; R =
CORRELATION COEF; P = PROBABILITY VALUE)
? T
ENTER T-TEST FOR DATA SET # 2
? 3.46
ENTER DF ASSOCIATED WITH THIS T-TEST
? 58
ENTER DIRECTION OF EFFECT ('+' IF IN EXPECTED
DIRECTION, '-' IF IN OPPOSITE DIRECTION)
? +

ENTER WEIGHT? 60

EFFECT SIZE STATISTICS FOR DATA SET # 2
CORRELATION COEFFICIENT:
          R= .413633
PROPORTION OF VARIANCE ACCOUNTED FOR:
        R[2= .171092
BINOMIAL EFFECT SIZE DISPLAY:   FROM
.293183   TO   .706817

STANDARDIZED DIFFERENCE BETWEEN MEANS:
          D= .908641

ENTER TO CONTINUE?

DATA SET # 3
ENTER TYPE OF STATISTIC FOR DATA SET # 3
(T = T-TEST; F = F-TEST; X = CHI-SQUARE; R =
CORRELATION COEF; P = PROBABILITY VALUE)
? F
ENTER F-TEST FOR DATA SET # 3
? 4.18
ENTER DF(ERROR) ASSOCIATED WITH THIS F-TEST
```

```
? 29
ENTER DIRECTION OF EFFECT ('+' IF IN EXPECTED
DIRECTION, '-' IF IN OPPOSITE DIRECTION)
? +

ENTER WEIGHT? 31

EFFECT SIZE STATISTICS FOR DATA SET # 3
CORRELATION COEFFICIENT:
        R= .354936
PROPORTION OF VARIANCE ACCOUNTED FOR:
      R[2= .125979
BINOMIAL EFFECT SIZE DISPLAY:   FROM
.322532   TO   .677468

STANDARDIZED DIFFERENCE BETWEEN MEANS:
        D= .75931

ENTER TO CONTINUE?

DATA SET # 4
ENTER TYPE OF STATISTIC FOR DATA SET # 4
(T = T-TEST; F = F-TEST; X = CHI-SQUARE; R =
CORRELATION COEF; P = PROBABILITY VALUE)
? R
ENTER CORRELATION COEFFICIENT FOR DATA SET #
4
? .5368
ENTER DF ASSOCIATED WITH THIS CORRELATION
COEFFICIENT
? 20
ENTER DIRECTION OF EFFECT ('+' IF IN EXPECTED
DIRECTION, '-' IF IN OPPOSITE DIRECTION)
? +

ENTER WEIGHT? 22

EFFECT SIZE STATISTICS FOR DATA SET # 4
CORRELATION COEFFICIENT:
        R= .5368
PROPORTION OF VARIANCE ACCOUNTED FOR:
      R[2= .288154
BINOMIAL EFFECT SIZE DISPLAY:   FROM
.2316   TO   .7684
```

```
STANDARDIZED DIFFERENCE BETWEEN MEANS:
        D= 1.27248

ENTER TO CONTINUE?

DATA SET # 5
ENTER TYPE OF STATISTIC FOR DATA SET # 5
(T = T-TEST; F = F-TEST; X = CHI-SQUARE; R =
CORRELATION COEF; P = PROBABILITY VALUE)
? P
ENTER PROBABILITY VALUE FOR DATA SET # 5
? .389
ENTER N ASSOCIATED WITH THIS PROBABILITY
VALUE
? 15
ENTER DIRECTION OF EFFECT ('+' IF IN EXPECTED
DIRECTION, '-' IF IN OPPOSITE DIRECTION)
? +

ENTER WEIGHT? 15

EFFECT SIZE STATISTICS FOR DATA SET # 5
CORRELATION COEFFICIENT:
        R= .0726826
PROPORTION OF VARIANCE ACCOUNTED FOR:
        R[2= 5.28277E-03
BINOMIAL EFFECT SIZE DISPLAY:  FROM
.463659  TO  .536342

STANDARDIZED DIFFERENCE BETWEEN MEANS:
        D= .145751

ENTER TO CONTINUE?
```

Note that the weights entered here are the sample sizes associated with each study. The program then displays the mean indices of effect size, indicating the combined effect size for the included studies. For example:

```
* * * META-ANALYTIC STATISTICS
FISHER Z FOR COMBINATION:
        Z= .34447
MEAN CORRELATION COEFFICIENT:
        R= .331462
```

```
PROPORTION OF VARIANCE ACCOUNTED FOR:
      R[2= .109867
BINOMIAL EFFECT SIZE DISPLAY:  FROM .334269
TO  .665731
MEAN STANDARDIZED DIFFERENCE BETWEEN
MEANS:
        D= .702646
```

Performing the computation of the Z for combination by hand:

$$Z = \frac{\Sigma wjZj}{\Sigma\, wj} = \frac{100\,(.26352) + 60\,(.439986) + 31\,(.37108) + 22\,(.599649) + 15\,(.0728108)}{100 + 60 + 31 + 22 + 15}$$

$$= .34447$$

Thus, the combined effect size obtained from the five studies listed in Table 4.1 appears to be moderate in magnitude.

Program L. Correlation Matrix, Partial Correlation and Multiple Correlation. This program requires the user to enter meta-analytic statistics (e.g., significance levels, effect sizes) and study attributes (e.g., duration of independent variable manipulation, date of publication, etc.). The user first enters the number of studies (rows) and the number of variables (columns). Next the user enters names for the variables to be entered (names should be restricted to brief, four-letter acronyms for the sake of output display; names like "SIGN", "EFFR", and "DATE" should be sufficient). The user then enters the relevant data for each study *in the order in which the variables were named.* For example:

```
CORRELATION, PARTIAL CORR. AND MULTIPLE
CORR.

NUMBER OF STUDIES (ROWS)? 5

NUMBER OF VARIABLES (COLUMNS)? 4
ENTER NAME OF VARIABLE # 1
? SIGN
ENTER NAME OF VARIABLE # 2
? EFFR
ENTER NAME OF VARIABLE # 3
? DATE
```

```
ENTER NAME OF VARIABLE # 4
? DURX
INPUT DATA FOR STUDY # 1
? .005
? .257585
? 1980
? 8
INPUT DATA FOR STUDY # 2
? .0005
? .413633
? 1980
? 10
INPUT DATA FOR STUDY # 3
? .025
? .354936
? 1981
? 4
INPUT DATA FOR STUDY #4
? .005
? .5368
? 1982
? 6
INPUT DATA FOR STUDY # 5
? .389
? .0726826
? 1983
? 2
```

The program then displays the means and the standard deviations for each variable, followed by a row-by-row listing of the matrix of correlations between all variables. For example:

```
MEAN SIGN = .0849          MEAN EFFR  = .327127
MEAN DATE = 1981.2         MEAN DURX  = 6

(ENTER TO CONTINUE)?
S.D. SIGN   = .170262      S.D. EFFR  = .174498
S.D. DATE   = 1.30384      S.D. DURX  = 3.16228

(ENTER TO CONTINUE)?
```

```
PRODUCT MOMENT CORRELATION MATRIX
SIGN:SIGN 1
SIGN:EFFR-  .82076
SIGN:DATE   .779976
SIGN:DURX-  .740136

(ENTER TO CONTINUE)?
EFFR:SIGN-  .82076
EFFR:EFFR 1
EFFR:DATE-  .347425
EFFR:DURX   .529666

(ENTER TO CONTINUE)?
DATE:SIGN   .779976
DATE:EFFR-  .347425
DATE:DATE 1
DATE:DURX-  .848875

(ENTER TO CONTINUE)?
DURX:SIGN-  .740136
DURX:EFFR   .529666
DURX:DATE-  .848875
DURX:DURX 1

(ENTER TO CONTINUE)?
```

Note that the correlation between significance levels and "Duration of X used in the study" ($r = -.740136$), while of very large magnitude, is not statistically significant ($p = .43365$). This might suggest that significance level is not significantly related to duration of X used in the study, whereas the focused comparison of Program H was able to substantiate this trend at the significance level of $p = .0182149$.

After initial display of the correlation matrix, the user is prompted to select computation of a partial correlation coefficient, computation of a multiple correlation coefficient, or re-display of the original correlation matrix.

If the partial correlation option is selected, the user is prompted to enter the names of the first and second variables (i.e., the variables of primary concern), and, the name of the variable whose effect is to be "partialed out" of the relationship between the other two. For example, suppose the user wanted to examine the relationship between effect size and duration of X used in the study, partialing out the influence of date of publication:

```
ENTER 'P' TO COMPUTE PARTIAL CORRELATION
ENTER 'M' TO COMPUTE MULTIPLE CORRELATION
ENTER 'R' TO DISPLAY CORRELATION MATRIX? P
PARTIAL CORRELATION

ENTER NAME OF FIRST VARIABLE? EFFR

ENTER NAME OF SECOND VARIABLE? DURX

ENTER NAME OF VARIABLE TO BE PARTIALED OUT?
DATE
PARTIAL CORRELATION COEFFICIENT
R(EFFR, DURX.DATE) = .473597
```

If the multiple correlation option is selected, the user is prompted to enter the name of the "criterion" variable (i.e., the variable of primary concern), and, the names of the two variables to be used to "predict" the criterion variable. For example, suppose the user wanted to examine the multiple correlation between duration of X used in the study, date of publication, and effect size, with duration of X and date of publication serving as predictors of effect size:

```
ENTER 'P' TO COMPUTE PARTIAL CORRELATION
ENTER 'M' TO COMPUTE MULTIPLE CORRELATION
ENTER 'R' TO DISPLAY CORRELATION MATRIX? M
MULTIPLE CORRELATION

ENTER NAME OF 'CRITERION' VARIABLE? EFFR

ENTER NAME OF FIRST 'PREDICTOR' VARIABLE?
DURX

ENTER NAME OF SECOND 'PREDICTOR' VARIABLE?
DATE
MULTIPLE CORRELATION COEFFICIENT
R (EFFR.DURX,DATE) = .563848
```

After computation of a partial correlation coefficient or a multiple correlation coefficient, the program returns the user to the choice between partial correlation, multiple correlation, or re-display of the original correlation matrix.

Program M. Disjoint Cluster Analysis of Effect Sizes. This program prompts the user to enter the number of studies to be included, and the effect size *(r)* and the associated N for each included study. For example:

```
DISJOINT CLUSTER ANALYSIS OF EFFECT SIZES (R)

ENTER NUMBER OF STUDIES TO BE INCLUDED? 5
ENTER EFFECT SIZE (R) FOR DATA SET # 1
? .257585
ENTER SAMPLE SIZE FOR DATA SET # 1
? 100
ENTER EFFECT SIZE (R) FOR DATA SET # 2
? .413633
ENTER SAMPLE SIZE FOR DATA SET # 2
? 60
ENTER EFFECT SIZE (R) FOR DATA SET # 3
? .354936
ENTER SAMPLE SIZE FOR DATA SET # 3
? 31
ENTER EFFECT SIZE (R) FOR DATA SET # 4
? .5368
ENTER SAMPLE SIZE FOR DATA SET # 4
? 22
ENTER EFFECT SIZE (R) FOR DATA SET # 5
? .0726826
ENTER SAMPLE SIZE FOR DATA SET # 5
? 15
```

After entering the r and N for each study, the user is then prompted to select a level of statistical significance (p = .10, .05 or .01). The program then displays the clusters that are obtained for that level of statistical significance. Within each cluster, studies are listed in rank order from highest effect size to lowest effect size. For example:

```
ENTER LEVEL OF SIGNIFICANCE FOR CLUSTER
ANALYSIS
(.10, .05, .01)? .05

LISTING OF CLUSTERS AT P = .05 LEVEL OF SIGNIFI-
CANCE
CLUSTER 1
     DATA SET # 4 (R = .5368)
     DATA SET # 2 (R = .413633)
     DATA SET # 3 (R = .354936)
     DATA SET # 1 (R = .257585)
     DATA SET # 5 (R = .0726826)
```

ENTER TO CONTINUE?

ENTER LEVEL OF SIGNIFICANCE FOR CLUSTER ANALYSIS
(.10, .05, .01)?

Performing the computation of this cluster analysis by hand first requires the computation of the square-mean-root estimate of common sample size:

$$\frac{\Sigma\ \sqrt{Nj-3}}{k} = \frac{\sqrt{(100-3)}+\sqrt{(60-3)}+\sqrt{(31-3)}+\sqrt{(22-3)}+\sqrt{(15-3)}}{5} = 6.10264$$

Then, the standard normal variate (Uj) can be computed for each study's effect size:

$$Uj = \left(\frac{\Sigma\sqrt{Nj-3}}{k}\right) Zj:$$

Williams (1982)	$U1$	= (6.10264) (.599649)	= 3.659442
Jones (1980)	$U2$	= (6.10264) (.439986)	= 2.685076
Johnson (1981)	$U3$	= (6.10264) (.37108)	= 2.264568
Smith (1980)	$U4$	= (6.10264) (.26352)	= 1.608168
Brown (1983)	$U5$	= (6.10264) (.0728108)	= .444338

Then the difference between adjacently-ranked Us are obtained and examined:

$$U1 - U2 = 3.659442 - 2.685076 = .974366$$
$$U2 - U3 = 2.685076 - 2.264568 = .420508$$
$$U3 - U4 = 2.264568 - 1.608168 = .656400$$
$$U4 - U5 = 1.608168 - .444338 = 1.16383$$

The critical value for a set of five studies at the significance level of $p = .05$ is 2.13 (Hedges & Olkin, 1983, Table 1). None of the differences between adjacently-ranked clusters are equal to or greater than this critical value; hence, the effect sizes of the five studies listed in Table 4.1 fall into a single "cluster." This is consistent with the nonsignificant diffuse test comparison of effect sizes illustrated above (Program F), which indicated that the effect sizes of the studies listed in Table 4.1 do not appear to be significantly heterogeneous.

Program N. Model Estimation. A different set of studies is used to illustrate the model estimation program. This new set of studies is listed in Table 4.2. These three studies yield a total of five separate comparisons between the dependent variable observed for smaller exemplars of the independent variable and the dependent variable observed for larger exemplars of the independent variable. A larger effect size *(d)* indicates that higher dependent variable measurements were obtained under conditions of lower

TABLE 4.2
Set of three studies examining the effects of varying amounts of *X* on *Y*

	Independent Variable		
Study (Date)	*Numerical Value of Smaller* X	*Numerical Value of Larger* X	*Derived Effect Size* (d)*
Smith (1981)	1	5	.43
Jones (1982)	1	12	1.20
Brown (1983)	1	5	.38
	1	10	1.09
	5	10	.29

*Note: larger effect size indicates that higher dependent variable measurements were obtained under conditions of lower levels of the independent variable (e.g., higher achievement obtained with smaller class sizes (Glass, 1980)).

levels of the independent variable (e.g., higher achievement scores obtained with smaller class sizes [Glass, 1980]).

The program prompts the user to enter the number of comparisons that will be included in the model estimation procedure, and also to enter the name of the independent variable (e.g., "X", "CLASS SIZE", "DURATION OF X", etc.). Then for each comparison, the program prompts the user to enter the numerical value of the smaller independent variable exemplar, the numerical value of the larger independent variable exemplar, and the study outcome (e.g., *r* or *d*). For example:

```
MODEL ESTIMATION

ENTER NUMBER OF COMPARISONS? 5

ENTER NAME OF INDEPENDENT VARIABLE? IV

COMPARISON # 1
ENTER NUMERICAL VALUE FOR SMALLER IV
? 1
ENTER NUMERICAL VALUE FOR LARGER IV
? 5
ENTER STUDY OUTCOME
? .43

COMPARISON # 2
ENTER NUMERICAL VALUE FOR SMALLER IV
? 1
ENTER NUMERICAL VALUE FOR LARGER IV
? 12
```

```
ENTER STUDY OUTCOME
? 1.2

COMPARISON # 3
ENTER NUMERICAL VALUE FOR SMALLER IV
? 1
ENTER NUMERICAL VALUE FOR LARGER IV
? 5
ENTER STUDY OUTCOME
? .38

COMPARISON # 4
ENTER NUMERICAL VALUE FOR SMALLER IV
? 1
ENTER NUMERICAL VALUE FOR LARGER IV
? 10
ENTER STUDY OUTCOME
? 1.09

COMPARISON # 5
ENTER NUMERICAL VALUE FOR SMALLER IV
? 5
ENTER NUMERICAL VALUE FOR LARGER IV
? 10
ENTER STUDY OUTCOME
? .29
```

The user is then prompted to specify the nature of the model to be estimated. This specification takes the form of an arithmetic operation performed on the independent variable values of the two exemplars for each comparison. Options which the user can choose are: (1) the ratio of the larger exemplar to the smaller exemplar; (2) the natural logarithm of the ratio of the larger exemplar to the smaller exemplar; and (3) exponentiation of the ratio of the larger exemplar to the smaller exemplar. If this final option is chosen, the program then prompts the user to supply the exponent to which the ratio will be raised. For example:

```
MODEL SPECIFICATION
ENTER  1 = (LARGER IV / SMALLER IV)
       2 = LOG (LARGER IV / SMALLER IV)
       3 = (LARGER IV / SMALLER IV) [ X
             (USER SUPPLIES THE EXPONENT 'X')
  ? 2
```

Note that in this example, the model specified is the natural logarithm of the ratio of the larger exemplar to the smaller exemplar. This program then performs a regression analysis, regressing the effect size for each comparison on the predictor formed by the selected model specification. After displaying the results of the regression analysis, the user is allowed to specify a different model, to be used in another regression analysis. For example:

```
* * * TEST OF MODEL FIT
MODEL: LOG (LARGER IV / SMALLER IV)
MEAN IV = 1.7399
S.D. IV = .632674
MEAN STUDY OUTCOME = .678
S.D. STUDY OUTCOME = .385507
COEFFICIENT OF REGRESSION (R) = .900442
COEFFICIENT OF DETERMINATION (R[2) = .810796

FORMULA FOR BEST-FITTING LINE:
        Y= -.276627+(.548667 X)
F-TEST FOR SIGNIFICANCE OF REGRESSION:
   F(1, 3)= 12.8559
ASSOCIATED ONE-TAILED P = .0354

ENTER TO CONDUCT ALTERNATIVE ANALYSIS?
```

Thus, the logarithm of the ratio of the larger exemplar to the smaller exemplar appears to provide a reasonably accurate model of the relationship between the independent variable and the dependent variable for the studies listed in Table 4.2.

One can produce a graphic illustration of the relation between the independent variable and the dependent variable defined by the model through the use of the formula for the least-squares regression line produced by the regression analysis illustrated above. This is done by holding the size of the smaller exemplar constant at "1", and then computing the predicted effect size $(\hat{d})$ for each larger exemplar across the range of observed values of the independent variable. For example, the predicted effect size for a larger exemplar of 5 and a smaller exemplar of 1 is: $\hat{d} = -.276627 + .548667(\log(5/1)) = .6064$; the predicted effect size for a larger exemplar of 10 and a smaller exemplar of 1 is; $\hat{d} = -.276627 + .548667(\log(10/1)) = .9867$. In this manner, the predicted effect sizes for each larger exemplar across the range of observed values of the independent variable, holding the smaller exemplar constant at "1", can be computed. These predicted effect sizes are illustrated in Figure 4.1; this figure serves as a graphic illustration

FIG. 4.1. Graphic illustration of the relation between the independent variable and the dependent variable of studies listed in Table 4.2. This is based on the model estimated $\hat{d} = -.276627 + .548667(\log(\text{larger exemplar/smaller exemplar}))$.

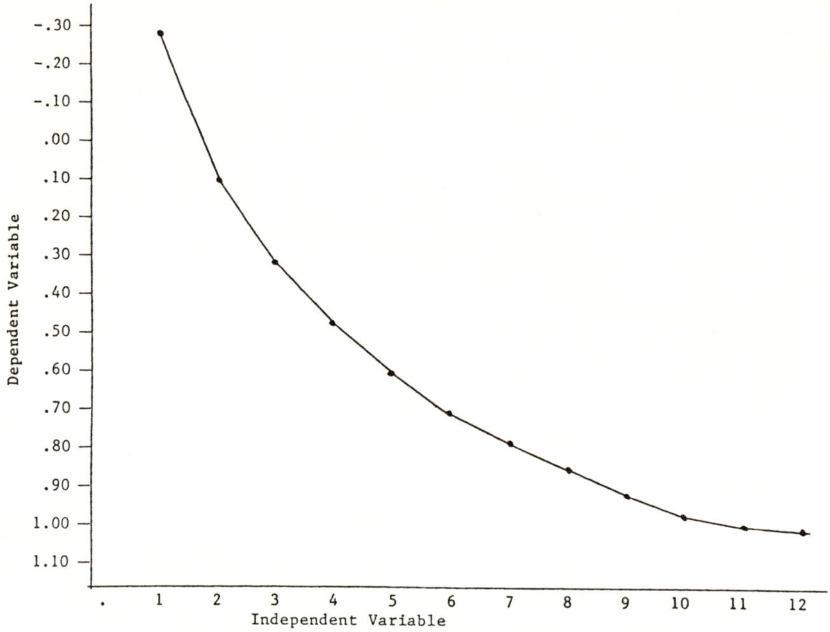

of the relation between the independent variable and the dependent variable as specified by the model (sp., the model defined by the natural logarithm of the ratio of the larger exemplar to the smaller exemplar).

Figure 4.1 indicates that at low levels of the independent variable, increases in the independent variable will produce substantial increases in effect size (i.e., substantial decreases in the dependent variable); whereas, at high levels of the independent variable, increases in the independent variable will produce slight increases in effect size (i.e., slight decreases in the dependent variable). This is similar to Glass's (1980) treatment of the effects of class size on achievement. For instance, with a class of relatively small size (e.g., one student), increasing the class by one student would be expected to produce a substantial decrease in achievement; however, with a class of relatively larger size (e.g., sixty students), increasing the class by one student would be expected to produce only a slight decrease in achievement (if any).

Limitations. The only p values that can be legitimately entered into these programs are one-tailed ps. The only χ^2 values that can be legitimately entered into these programs are those χ^2 values with one degree of freedom. Similarly, the only F test values that can be legitimately entered into these programs are those for which the degrees of freedom in the numerator is equal to one, as in the comparison between two group means or the test for a trend or other contrast.

These programs can accept p values between $p = .99999996$ ($Z = -5.2944$) and $p = .0000000000000000001$ ($Z = 9.0129$). If the user should include in their meta-analysis a study that produced a p value more extreme than these limits, the most extreme value the program will accept (indicated above) can serve as a conservative estimate of the real p. There are no limits on the size of the t, F, or r statistics that can be entered into these programs. The significance level programs can also accept a χ^2 of any size, but the effect sizes programs may produce an error if a χ^2 value is entered which is larger than the N associated with this χ^2 (the Pearson r derived for a χ^2 of this magnitude is greater than 1.00). Such an extreme χ^2 is most likely to occur with a small sample size, and with an expected frequency substantially less than five. If a user includes a study that produced a χ^2 larger than the N associated with it, setting the χ^2 equal to the associated N can provide a more conservative estimate.

The values retrieved by these programs from the entered study data are generally accurate to within ± .005 of the values obtained from standard statistical tables and/or from computational formulae. The most extreme p value that these programs will retrieve for a computed Z for comparison or combination is $p = .000000001$ ($Z = 6.03788$); more extreme Zs may be computed, but in such cases the associated p will be correctly, but imprecisely, reported as "$p<.000000001$. This limitation is imposed upon the user by the number of significant digits which computer systems typically can handle for internal computations, and by the imprecision of the algorithm used beyond this point. For more extreme computed Zs for comparison or combination, the user may prefer to look up the associated p in an extended table of the normal probability distribution (e.g., Abramowitz & Stegun, 1964). Similarly, these programs may be unable to compute an exact p value for extreme computed χ^2 for comparison (there is not a single given limit to this calculation, since the p values associated with these computed χ^2s depend on the degrees of freedom associated with the χ^2 in question). If the program is unable to retrieve an exact p value for an extreme computed χ^2 for comparison, the associated p will be correctly, but imprecisely, reported as "$p<.0000001$". Again, the user may prefer to look up the p associated with such extreme computed χ^2s in an extended table of chi-square probability distribution (e.g., Abramowitz & Stegun, 1964).

The number of studies which can be included in an analysis using these computer programs is determined by the dimension commands in the beginning of each program. The number within parentheses following a DIM command defines the maximum number of studies the program is expected to handle. For example, line 00060 of Program A defines a vector ZM as having room for two studies with the command:

```
00060 DIM ZM (2)
```

Programs A through D can each accommodate two studies. Programs E, F, H, I, J, K and N can each accommodate 500 studies. Program G can accommodate 450 levels of the study attribute. Program L can accommodate 100 studies and eight variables. Program M can accommodate 300 studies. A computer system with very limited random access memory (RAM) might not have the memory capacity to support these large DIM commands.

Reducing the number within the parentheses following a program's DIM command(s) will cause these programs to require less RAM, and thereby allow these programs to operate on computer systems with limited RAM capacity. For example, changing the DIM command on line 00150 of Program E from "DIM ZM (500)" to "DIM ZM (100)" would cause this program to require less RAM capacity. This change from "500" to "100" would allow the user to perform diffuse comparison of the significance levels of up to 100 studies.

5 Meta-analysis Programs

PROGRAM A. COMPARISON OF SIGNIFICANCE LEVELS (TWO STUDIES)

```
00010  PRINT" META-ANALYSIS COMPARING SIGNIFICANCE LEVELS"
00020 PRINT" (TWO STUDIES)"
00030 DE$="ENTER DIRECTION OF EFFECT ('+' IF EXPECTED, '-' IF UNEXPECTED
)"
00040 N=0
00050 ZM=0
00060 DIM ZM (2)
00070 N=N+1
00080 PRINT"ENTER TYPE OF STATISTIC FOR DATA SET #" ;N
00090 PRINT "(T=T-TEST; F=F-TEST; X=CHI-SQUARE; R=CORR COEF; P=PROB VALU
E)"
00100 INPUT A$
00110 IF A$="T" THEN GOTO 180
00120 IF A$="F" THEN GOTO 270
00130 IF A$="X" THEN GOTO 350
00140 IF A$="R" THEN GOTO 420
00150 IF A$="P" THEN GOTO 500
00160 N=N-1
00170 GOTO 70
00180 PRINT"ENTER T-TEST FOR DATA SET #";N
00190 INPUT S
00200 PRINT"ENTER DF ASSOCIATED WITH THIS T-TEST"
00210 INPUT DF
```

```
00220 PRINT DE$
00230 INPUT E$
00240 Z=(DF*LOG((1+(S**2/DF))))**.5*(1-(1/(2*DF)))**.5
00250 IF E$="-" THEN Z=-Z
00260 GOTO 590
00270 PRINT"ENTER F-TEST FOR DATA SET #";N
00280 INPUT S
00290 PRINT"ENTER DF(ERROR) ASSOCIATED WITH THIS F-TEST"
00300 INPUT DF
00310 PRINT DE$
00320 INPUT E$
00330 S=SQR(S)
00340 GOTO 240
00350 PRINT"ENTER CHI-SQUARE FOR DATA SET #";N
00360 INPUT S
00370 PRINT DE$
00380 INPUT E$
00390 Z=SQR(S)
00400 IF E$="-" THEN Z=-Z
00410 GOTO 590
00420 PRINT"ENTER CORRELATION COEFFICIENT FOR DATA SET #";N
00430 INPUT S
00440 PRINT"ENTER DF ASSOCIATED WITH THIS CORRELATION"
00450 INPUT DF
00460 PRINT DE$
00470 INPUT E$
```

(Continued)

PROGRAM A *(Continued)*

```
00480 S=(S*SQR(DF))/SQR(1-S**2)
00490 GOTO 240
00500 PRINT"ENTER ONE-TAILED P VALUE FOR DATA SET #";N
00510 INPUT P
00520 IF P >.5 THEN X=1
00530 IF P > .5 THEN P=1-P
00540 Q=SQR(LOG(1/(P**2)))
00550 Z1=(2.515517+.802853*Q+.010328*(Q**2))
00560 Z2=(1+1.432788*Q+.189269*(Q**2)+.001308*(Q**3))
00570 Z=Q-(Z1/Z2)
00580 IF X=1 THEN Z=-Z
00590 PRINT"Z = ";Z
00600 ZM(N)=Z
00610 Z=0
00620 X=0
00630 E$=""
00640 IF N < 2 THEN GOTO 70
00650 ZM = ABS(ZM(1) - ZM(2))/SQR(2)
00660 PRINT"* * * META-ANALYTIC STATISTICS"
00670 PRINT"Z FOR COMPARISON = ";ZM
00680 P1=(1+.049867347*ZM+.0211410061*(ZM**2)+.0032776263*(ZM**3))
00690 P2=(.0000380036*(ZM**4)+.0000488906*(ZM**5)+.000005383*(ZM**6))
00700 P=.5*(P1+P2)**-16
00710 IF ZM < 6.03788 THEN PRINT"ASSOCIATED ONE-TAILED P = ";P
00720 IF ZM >= 6.03788 THEN PRINT"ASSOCIATED ONE-TAILED P < .000000001"
00730 END
```

PROGRAM B. COMPARISON OF EFFECT SIZES (TWO STUDIES)

```
00010 PRINT " META-ANALYSIS COMPARING EFFECT SIZES"
00020 PRINT " (TWO STUDIES)"
00030 DE$="ENTER DIRECTION OF EFFECT ('+' IF EXPECTED, '-' IF UNEXPECTED
)"
00040 N=0
00050 R=0
00060 ZN=0
00070 DIM AR (2), NR (2), D (2)
00080 N=N+1
00090 E$=""
00100 PRINT "DATA SET #";N
00110 PRINT "ENTER TYPE OF STATISTIC FOR DATA SET # ";N
00120 PRINT "(T=T-TEST; F=F-TEST; X=CHI-SQUARE; R=CORR COEF; P=PROB VALU
E)"
00130 INPUT A$
00140 IF A$="T" THEN GOTO 200
00150 IF A$="F" THEN GOTO 310
00160 IF A$="X" THEN GOTO 420
00170 IF A$="R" THEN GOTO 690
00180 IF A$="P" THEN GOTO 540
00190 GOTO 100
00200 PRINT "ENTER T-TEST FOR DATA SET#";N
00210 INPUT S
```

(Continued)

PROGRAM B *(Continued)*

```
00220 PRINT "ENTER DF ASSOCIATED WITH THIS T-TEST"
00230 INPUT DF
00240 PRINT DE$
00250 INPUT E$
00260 R=SQR(S**2/(S**2+DF))
00270 IF E$="-" THEN R=-R
00280 D=SQR((4*R**2)/(1-R**2))
00290 IF E$="-" THEN D=-D
00300 GOTO 790
00310 PRINT "ENTER F-TEST FOR DATA SET #";N
00320 INPUT S
00330 PRINT "ENTER DF(ERROR) ASSOCIATED WITH THIS F-TEST"
00340 INPUT DF
00350 PRINT DE$
00360 INPUT E$
00370 R=SQR(S/(S+DF))
00380 IF E$="-" THEN R=-R
00390 D=SQR((4*R**2)/(1-R**2))
00400 IF E$="-" THEN D=-D
00410 GOTO 790
00420 PRINT "ENTER CHI-SQUARE FOR DATA SET #";N
00430 INPUT S
00440 PRINT "ENTER N ASSOCIATED WITH THIS CHI-SQUARE"
00450 INPUT DF
```

```
00460 PRINT DE$
00470 INPUT E$
00480 R=SQR(S/DF)
00490 IF E$="-" THEN R=-R
00500 D=SQR((4*R**2)/(1-R**2))
00510 IF E$="-" THEN D=-D
00520 DF=DF-2
00530 GOTO 790
00540 PRINT "ENTER ONE-TAILED P VALUE FOR DATA SET # ";N
00550 INPUT S
00560 PRINT "ENTER N ASSOCIATED WITH THIS PROBABILITY VALUE"
00570 INPUT DF
00580 PRINT DE$
00590 INPUT E$
00600 IF S >.5 THEN S=1-S
00610 Q=SQR(LOG(1/(S**2)))
00620 Z1=(2.515517+.802853*Q+.010328*(Q**2))
00630 Z2=(1+1.432788*Q+.189269*(Q**2)+.001308*(Q**3))
00640 Z=Q-(Z1/Z2)
00650 R=SQR(Z**2/DF)
00660 IF E$="-" THEN R=-(ABS(R))
00670 DF=DF-2
00680 GOTO 770
00690 PRINT "ENTER CORRELATION COEFFICIENT FOR DATA SET #";N
00700 INPUT S
```

(Continued)

PROGRAM B *(Continued)*

```
00710 PRINT "ENTER DF ASSOCIATED WITH THIS CORRELATION COEFFICIENT"
00720 INPUT DF
00730 PRINT DE$
00740 INPUT E$
00750 R=S
00760 IF E$="-" THEN R=-(ABS(R))
00770 D=SQR((4*R**2)/(1-R**2))
00780 IF E$="-" THEN D=-D
00790 ZR=((LOG(1+R))-(LOG(1-R)))/2
00800 AR(N)=ZR
00810 NR(N)=DF-1
00820 B=.5+(R/2)
00830 PRINT "EFFECT SIZE STATISTICS FOR DATA SET #";N
00840 PRINT "CORRELATION COEFFICIENT:           R = ";R
00850 PRINT "PROPORTION OF VARIANCE ACCOUNTED FOR:  R2 = ";R**2
00860 PRINT "BINOMIAL EFFECT SIZE DISPLAY:    FROM ";1-B;"TO ";B
00870 PRINT "STANDARDIZED DIFFERENCE BETWEEN MEANS:   D = ";D
00880 X=0
00890 R=0
00900 Z=0
00910 D=0
00920 PRINT "ENTER TO CONTINUE"
00930 INPUT A$
00940 IF N < 2  THEN GOTO 80
```

```
00950 PRINT " * * * META-ANALYTIC STATISTICS"
00960 Z =ABS(AR(1) - AR(2))/SQR(1/NR(1) + 1/NR(2))
00970 PRINT"Z FOR COMPARISON = ";Z
00980 P1=(1+.049867347*Z+.0211410061*(Z**2)+.0032776263*(Z**3))
00990 P2=(.0000380036*(Z**4)+.0000488906*(Z**5)+.000005383*(Z**6))
01000 P=.5*(P1+P2)**-16
01010 IF Z < 6.03788 THEN PRINT"ASSOCIATED ONE-TAILED P = ";P
01020 IF Z >= 6.03788 THEN PRINT"ASSOCIATED ONE-TAILED P < .000000001"
01030 END
```

PROGRAM C. COMBINATION OF SIGNIFICANCE LEVELS (TWO STUDIES)

```
00010  PRINT" META-ANALYSIS COMBINING SIGNIFICANCE LEVELS"
00020 PRINT" (TWO STUDIES)"
00030 DE$="ENTER DIRECTION OF EFFECT ('+' IF EXPECTED, '-' IF UNEXPECTED
)"
00040 N=0
00050 ZM=0
00060 ZD=0
00070 DIM ZM (2)
00080 N=N+1
00090 PRINT"ENTER TYPE OF STATISTIC FOR DATA SET #" ;N
00100 PRINT "(T=T-TEST; F=F-TEST; X=CHI-SQUARE; R=CORR COEF; P=PROB VALU
E)"
00110 INPUT A$
00120 IF A$="T" THEN GOTO 190
00130 IF A$="F" THEN GOTO 280
00140 IF A$="X" THEN GOTO 360
00150 IF A$="R" THEN GOTO 430
00160 IF A$="P" THEN GOTO 510
00170 N=N-1
00180 GOTO 80
00190 PRINT"ENTER T-TEST FOR DATA SET #";N
00200 INPUT S
```

```
00210 PRINT"ENTER DF ASSOCIATED WITH THIS T-TEST"
00220 INPUT DF
00230 PRINT DE$
00240 INPUT E$
00250 Z=(DF*LOG((1+(S**2/DF))))**.5*(1-(1/(2*DF)))**.5
00260 IF E$="-" THEN Z=-Z
00270 GOTO 600
00280 PRINT"ENTER F-TEST FOR DATA SET #";N
00290 INPUT S
00300 PRINT"ENTER DF(ERROR) ASSOCIATED WITH THIS F-TEST"
00310 INPUT DF
00320 PRINT DE$
00330 INPUT E$
00340 S=SQR(S)
00350 GOTO 250
00360 PRINT"ENTER CHI-SQUARE FOR DATA SET #";N
00370 INPUT S
00380 PRINT DE$
00390 INPUT E$
00400 Z=SQR(S)
00410 IF E$="-" THEN Z=-Z
00420 GOTO 600
00430 PRINT"ENTER CORRELATION COEFFICIENT FOR DATA SET #";N
00440 INPUT S
00450 PRINT"ENTER DF ASSOCIATED WITH THIS CORRELATION"
```

(Continued)

```
00460 INPUT DF
00470 PRINT DE$
00480 INPUT E$
00490 S=(S*SQR(DF))/SQR(1-S**2)
00500 GOTO 250
00510 PRINT"ENTER ONE-TAILED P VALUE FOR DATA SET #";N
00520 INPUT P
00530 IF P >.5 THEN X=1
00540 IF P > .5 THEN P=1-P
00550 Q=SQR(LOG(1/(P**2)))
00560 Z1=(2.515517+.802853*Q+.010328*(Q**2))
00570 Z2=(1+1.432788*Q+.189269*(Q**2)+.001308*(Q**3))
00580 Z=Q-(Z1/Z2)
00590 IF X=1 THEN Z=-Z
00600 PRINT"ENTER WEIGHT"
00610 INPUT W
00620 PRINT"Z = ";Z
00630 ZM=ZM+Z*W
00640 ZD=ZD+W**2
00650 Z=0
00660 W=0
00670 X=0
00680 E$=""
00690 IF N < 2 THEN GOTO 80
```

```
00700 ZM=ABS(ZM/SQR(ZD))
00710 PRINT"* * * META-ANALYTIC STATISTICS"
00720 PRINT"Z FOR COMBINATION = ";ZM
00730 Z1=(1+.049867347*ZM+.0211410061*(ZM**2)+.0032776263*(ZM**3))
00740 Z2=(.0000380036*(ZM**4)+.0000488906*(ZM**5)+.000005383*(ZM**6))
00750 P=.5*(Z1+Z2)**-16
00760 IF ZM < 6.03788 THEN PRINT"ASSOCIATED ONE-TAILED P = ";P
00770 IF ZM >= 6.03788 THEN PRINT"ASSOCIATED ONE-TAILED P<.000000001"
00780 END
```

PROGRAM D. COMBINATION OF EFFECT SIZES (TWO STUDIES)

```
00010 PRINT "    META-ANALYSIS COMBINING EFFECT SIZES"
00020 PRINT "    (TWO STUDIES)"
00030 DE$="ENTER DIRECTION OF EFFECT ('+' IF EXPECTED, '-' IF UNEXPECTED
)"
00040 N=0
00050 ZN=0
00060 W2=0
00070 DIM ZM (2)
00080 N=N+1
00090 E$=""
00100 PRINT "DATA SET #";N
00110 PRINT "ENTER TYPE OF STATISTIC FOR DATA SET # ";N
00120 PRINT "(T=T-TEST; F=F-TEST; X=CHI-SQUARE; R=CORR COEF; P=PROB VALU
E)"
00130 INPUT A$
00140 IF A$="T" THEN GOTO 200
00150 IF A$="F" THEN GOTO 310
00160 IF A$="X" THEN GOTO 420
00170 IF A$="R" THEN GOTO 690
00180 IF A$="P" THEN GOTO 540
00190 GOTO 100
00200 PRINT "ENTER T-TEST FOR DATA SET#";N
00210 INPUT S
```

```
00220 PRINT "ENTER DF ASSOCIATED WITH THIS T-TEST"
00230 INPUT DF
00240 PRINT DE$
00250 INPUT E$
00260 R=SQR(S**2/(S**2+DF))
00270 IF E$="-" THEN R=-R
00280 D=SQR((4*R**2)/(1-R**2))
00290 IF E$="-" THEN D=-D
00300 GOTO 790
00310 PRINT "ENTER F-TEST FOR DATA SET #";N
00320 INPUT S
00330 PRINT "ENTER DF(ERROR) ASSOCIATED WITH THIS F-TEST"
00340 INPUT DF
00350 PRINT DE$
00360 INPUT E$
00370 R=SQR(S/(S+DF))
00380 IF E$="-" THEN R=-R
00390 D=SQR((4*R**2)/(1-R**2))
00400 IF E$="-" THEN D=-D
00410 GOTO 790
00420 PRINT "ENTER CHI-SQUARE FOR DATA SET #";N
00430 INPUT S
00440 PRINT "ENTER N ASSOCIATED WITH THIS CHI-SQUARE"
00450 INPUT DF
00460 PRINT DE$
00470 INPUT E$
```

(Continued)

PROGRAM D *(Continued)*

```
00480 R=SQR(S/DF)
00490 IF E$="-" THEN R=-R
00500 D=SQR((4*R**2)/(1-R**2))
00510 IF E$="-" THEN D=-D
00520 DF=DF-2
00530 GOTO 790
00540 PRINT "ENTER ONE-TAILED P VALUE FOR DATA SET # ";N
00550 INPUT S
00560 PRINT "ENTER N ASSOCIATED WITH THIS PROBABILITY VALUE"
00570 INPUT DF
00580 PRINT DE$
00590 INPUT E$
00600 IF S >.5 THEN S=1-S
00610 Q=SQR(LOG(1/(S**2)))
00620 Z1=(2.515517+.802853*Q+.010328*(Q**2))
00630 Z2=(1+1.432788*Q+.189269*(Q**2)+.001308*(Q**3))
00640 Z=Q-(Z1/Z2)
00650 R=SQR(Z**2/DF)
00660 IF E$="-" THEN R=-(ABS(R))
00670 DF=DF-2
00680 GOTO 770
00690 PRINT "ENTER CORRELATION COEFFICIENT FOR DATA SET #";N
00700 INPUT S
00710 PRINT "ENTER DF ASSOCIATED WITH THIS CORRELATION COEFFICIENT"
00720 INPUT DF
```

```
00730 PRINT DE$
00740 INPUT E$
00750 R=S
00760 IF E$="-" THEN R=-(ABS(R))
00770 D=SQR((4*R**2)/(1-R**2))
00780 IF E$="-" THEN D=-D
00790 PRINT "ENTER WEIGHT"
00800 INPUT W1
00810 ZR=((LOG(1+R))-(LOG(1-R)))/2
00820 ZN =ZN+ZR*W1
00830 W2=W2+W1
00840 B=.5+(R/2)
00850 PRINT "EFFECT SIZE STATISTICS FOR DATA SET #";N
00860 PRINT "CORRELATION COEFFICIENT:          R = ";R
00870 PRINT "PROPORTION OF VARIANCE ACCOUNTED FOR:  R2 = ";R**2
00880 PRINT "BINOMIAL EFFECT SIZE DISPLAY:    FROM ";1-B;"TO ";B
00890 PRINT "STANDARDIZED DIFFERENCE BETWEEN MEANS:   D = ";D
00900 X=0
00910 R=0
00920 Z=0
00930 W1=0
00940 D=0
00950 PRINT "ENTER TO CONTINUE"
00960 INPUT A$
00970 IF N < 2  THEN GOTO 80
```

(Continued)

PROGRAM D *(Continued)*

```
00980 PRINT " * * * META-ANALYTIC STATISTICS"
00990 ZN=ZN/W2
01000 RN=((EXP(ZN))**2-1)/((EXP(ZN))**2+1)
01010 B=.5+(RN/2)
01020 D=SQR((4*RN**2)/(1-RN**2))
01030 PRINT "FISHER Z FOR COMBINATION:          Z = ";ZN
01040 PRINT "MEAN CORRELATION COEFFICIENT:          R = ";RN
01050 PRINT "PROPORTION OF VARIANCE ACCOUNTED FOR:  R2 = ";RN**2
01060 PRINT "BINOMIAL EFFECT SIZE DISPLAY:    FROM";1-B;" TO ";B
01070 PRINT "MEAN STANDARDIZED DIFFERENCE BETWEEN MEANS:  D = ";D
01080 END
```

PROGRAM E. COMPARISON (DIFFUSE) OF SIGNIFICANCE LEVELS (THREE OR MORE STUDIES)

```
00010  PRINT" META-ANALYSIS COMPARING SIGNIFICANCE LEVELS"
00020 PRINT" (DIFFUSE TEST)"
00030 PRINT" (THREE OR MORE STUDIES)"
00040 DE$="ENTER DIRECTION OF EFFECT ('+' IF EXPECTED, '-' IF UNEXPECTED
)"
00050 N=0
00060 ZM=0
00070 ZN=0
00080 X2=0
00090 DX=0
00100 R=0
00110 M=0
00120 LQ=0
00130 PRINT"ENTER THE NUMBER OF STUDIES TO BE INCLUDED"
00140 INPUT G
00150 DIM ZM (500)
00160 N=N+1
00170 PRINT"ENTER TYPE OF STATISTIC FOR DATA SET #" ;N
00180 PRINT "(T=T-TEST; F=F-TEST; X=CHI-SQUARE; R=CORR COEF; P=PROB VALU
E)"
00190 INPUT A$
00200 IF A$="T" THEN GOTO 270
00210 IF A$="F" THEN GOTO 360
```

(Continued)

PROGRAM E *(Continued)*

```
00220 IF A$="X" THEN GOTO 440
00230 IF A$="R" THEN GOTO 510
00240 IF A$="P" THEN GOTO 590
00250 N=N-1
00260 GOTO 160
00270 PRINT"ENTER T-TEST FOR DATA SET #";N
00280 INPUT S
00290 PRINT"ENTER DF ASSOCIATED WITH THIS T-TEST"
00300 INPUT DF
00310 PRINT DE$
00320 INPUT E$
00330 Z=(DF*LOG((1+(S**2/DF))))**.5*(1-(1/(2*DF)))**.5
00340 IF E$="-" THEN Z=-Z
00350 GOTO 680
00360 PRINT"ENTER F-TEST FOR DATA SET #";N
00370 INPUT S
00380 PRINT"ENTER DF(ERROR) ASSOCIATED WITH THIS F-TEST"
00390 INPUT DF
00400 PRINT DE$
00410 INPUT E$
00420 S=SQR(S)
00430 GOTO 330
00440 PRINT"ENTER CHI-SQUARE FOR DATA SET #";N
00450 INPUT S
```

```
00460 PRINT DE$
00470 INPUT E$
00480 Z=SQR(S)
00490 IF E$="-" THEN Z=-Z
00500 GOTO 680
00510 PRINT"ENTER CORRELATION COEFFICIENT FOR DATA SET #";N
00520 INPUT S
00530 PRINT"ENTER DF ASSOCIATED WITH THIS CORRELATION"
00540 INPUT DF
00550 PRINT DE$
00560 INPUT E$
00570 S=(S*SQR(DF))/SQR(1-S**2)
00580 GOTO 330
00590 PRINT"ENTER ONE-TAILED P VALUE FOR DATA SET #";N
00600 INPUT P
00610 IF P >.5 THEN X=1
00620 IF P > .5 THEN P=1-P
00630 Q=SQR(LOG(1/(P**2)))
00640 Z1=(2.515517+.802853*Q+.010328*(Q**2))
00650 Z2=(1+1.432788*Q+.189269*(Q**2)+.001308*(Q**3))
00660 Z=Q-(Z1/Z2)
00670 IF X=1 THEN Z=-Z
00680 PRINT"Z = ";Z
00690 ZN=ZN+Z
00700 ZM(N)=Z
```

(Continued)

```
00710 Z=0
00720 X=0
00730 E$=""
00740 IF N < G THEN GOTO 160
00750 ZN=ZN/N
00760 FOR X=1 TO N
00770 X2=X2+(ZM(X)-ZN)**2
00780 NEXT X
00790 DF=N-1
00800 PRINT"* * * META-ANALYTIC STATISTICS"
00810 PRINT"CHI-SQUARE FOR COMPARISON = ";X2
00820 PRINT"DF = ";DF
00830 IF X2=0 THEN GOTO 1050
00840 R=1
00850 FOR I = DF TO 2 STEP -2
00860 R=R*I
00870 NEXT I
00880 KQ=X2**(INT((DF+1)/2))*EXP(-X2/2)/R
00890 IF INT(DF/2)=DF/2 THEN J=1
00900 IF INT(DF/2)=DF/2 THEN GOTO 920
00910 J=SQR(2/X2/3.1415927)
00920 LQ=1
00930 M=1
00940 DX=DF
00950 DX=DX+2
```

```
00960 M=M*X2/DX
00970 IF M < .0000001 THEN GOTO 1000
00980 LQ=LQ+M
00990 GOTO 950
01000 P=1-J*KQ*LQ
01010 IF P > 0 THEN PRINT"ASSOCIATED ONE-TAILED P = ";P
01020 IF P > 0 THEN GOTO 1060
01030 PRINT"ASSOCIATED ONE-TAILED P < .000000001"
01040 GOTO 1060
01050 PRINT"ASSOCIATED ONE-TAILED P = 1.00"
01060 END
```

PROGRAM F. COMPARISON (DIFFUSE) OF EFFECT SIZES (THREE OR MORE STUDIES)

```
00010 PRINT " META-ANALYSIS COMPARING EFFECT SIZES"
00020 PRINT" (DIFFUSE TEST)"
00030 PRINT " (THREE OR MORE STUDIES)"
00040 DE$="ENTER DIRECTION OF EFFECT ('+' IF EXPECTED, '-' IF UNEXPECTED
)"
00050 N=0
00060 X2=0
00070 DX=0
00080 R=0
00090 M=0
00100 LQ=0
00110 ZN=0
00120 PRINT "ENTER THE NUMBER OF STUDIES TO BE INCLUDED"
00130 INPUT G
00140 DIM AR (500), AD (500), NR (500), D (500), DW (500)
00150 N=N+1
00160 E$=""
00170 ER=0
00180 ED=0
00190 BR=0
00200 PRINT "DATA SET #";N
00210 PRINT "ENTER TYPE OF STATISTIC FOR DATA SET # ";N
00220 PRINT "(T=T-TEST; F=F-TEST; X=CHI-SQUARE; R=CORR COEF; P=PROB VALU
E)"
```

```
00230 INPUT A$
00240 IF A$="T" THEN GOTO 300
00250 IF A$="F" THEN GOTO 410
00260 IF A$="X" THEN GOTO 520
00270 IF A$="R" THEN GOTO 790
00280 IF A$="P" THEN GOTO 640
00290 GOTO 200
00300 PRINT "ENTER T-TEST FOR DATA SET#";N
00310 INPUT S
00320 PRINT "ENTER DF ASSOCIATED WITH THIS T-TEST"
00330 INPUT DF
00340 PRINT DE$
00350 INPUT E$
00360 R=SQR(S**2/(S**2+DF))
00370 IF E$="-" THEN R=-R
00380 D=SQR((4*R**2)/(1-R**2))
00390 IF E$="-" THEN D=-D
00400 GOTO 890
00410 PRINT "ENTER F-TEST FOR DATA SET #";N
00420 INPUT S
00430 PRINT "ENTER DF(ERROR) ASSOCIATED WITH THIS F-TEST"
00440 INPUT DF
00450 PRINT DE$
00460 INPUT E$
00470 R=SQR(S/(S+DF))
```

(Continued)

PROGRAM F *(Continued)*

```
00480 IF E$="-" THEN R=-R
00490 D=SQR((4*R**2)/(1-R**2))
00500 IF E$="-" THEN D=-D
00510 GOTO 890
00520 PRINT "ENTER CHI-SQUARE FOR DATA SET #";N
00530 INPUT S
00540 PRINT "ENTER N ASSOCIATED WITH THIS CHI-SQUARE"
00550 INPUT DF
00560 PRINT DE$
00570 INPUT E$
00580 R=SQR(S/DF)
00590 IF E$="-" THEN R=-R
00600 D=SQR((4*R**2)/(1-R**2))
00610 IF E$="-" THEN D=-D
00620 DF=DF-2
00630 GOTO 890
00640 PRINT "ENTER ONE-TAILED P VALUE FOR DATA SET # ";N
00650 INPUT S
00660 PRINT "ENTER N ASSOCIATED WITH THIS PROBABILITY VALUE"
00670 INPUT DF
00680 PRINT DE$
00690 INPUT E$
00700 IF S >.5 THEN S=1-S
00710 Q=SQR(LOG(1/(S**2)))
00720 Z1=(2.515517+.802853*Q+.010328*(Q**2))
```

```
00730 Z2=(1+1.432788*Q+.189269*(Q**2)+.001308*(Q**3))
00740 Z=Q-(Z1/Z2)
00750 R=SQR(Z**2/DF)
00760 IF E$="-" THEN R=-(ABS(R))
00770 DF=DF-2
00780 GOTO 870
00790 PRINT "ENTER CORRELATION COEFFICIENT FOR DATA SET #";N
00800 INPUT S
00810 PRINT "ENTER DF ASSOCIATED WITH THIS CORRELATION COEFFICIENT"
00820 INPUT DF
00830 PRINT DE$
00840 INPUT E$
00850 R=S
00860 IF E$="-" THEN R=-(ABS(R))
00870 D=SQR((4*R**2)/(1-R**2))
00880 IF E$="-" THEN D=-D
00890 ZR=((LOG(1+R))-(LOG(1-R)))/2
00900 AR(N)=ZR
00910 NR(N)=DF-1
00920 B=.5+(R/2)
00930 PRINT "EFFECT SIZE STATISTICS FOR DATA SET #";N
00940 PRINT "CORRELATION COEFFICIENT:            R = ";R
00950 PRINT "PROPORTION OF VARIANCE ACCOUNTED FOR:  R2 = ";R**2
00960 PRINT "BINOMIAL EFFECT SIZE DISPLAY:    FROM ";1-B;"TO ";B
00970 PRINT "STANDARDIZED DIFFERENCE BETWEEN MEANS:   D = ";D
```

(Continued)

PROGRAM F *(Continued)*

```
00980 X=0
00990 R=0
01000 Z=0
01010 D=0
01020 PRINT "ENTER TO CONTINUE"
01030 INPUT A$
01040 IF N < G  THEN GOTO 150
01050 FOR X=1 TO N
01060 ER=ER+(AR(X)*NR(X))
01070 BR=BR+NR(X)
01080 NEXT X
01090 ER=ER/BR
01100 PRINT " * * * META-ANALYTIC STATISTICS"
01110 FOR X=1 TO N
01120 X2=X2+NR(X)*(AR(X)-ER)**2
01130 NEXT X
01140 DF=N-1
01150 PRINT"CHI-SQUARE FOR COMPARISON = ";X2
01160 PRINT"DF = ";DF
01170 IF X2=0 THEN GOTO 1390
01180 R=1
01190 FOR I=DF TO 2 STEP -2
01200 R=R*I
01210 NEXT I
01220 KQ=X2**(INT((DF+1)/2))*EXP(-X2/2)/R
```

```
01230 IF INT(DF/2)=DF/2 THEN J=1
01240 IF INT(DF/2)=DF/2 THEN GOTO 1260
01250 J=SQR(2/X2/3.1415927)
01260 LQ=1
01270 M=1
01280 DX=DF
01290 DX=DX+2
01300 M=M*X2/DX
01310 IF M < .0000001 THEN GOTO 1340
01320 LQ=LQ+M
01330 GOTO 1290
01340 P=1-J*KQ*LQ
01350 IF P > 0 THEN PRINT"ASSOCIATED ONE-TAILED P = ";P
01360 IF P > 0 THEN GOTO 1400
01370 PRINT"ASSOCIATED ONE-TAILED P < .000000001"
01380 GOTO 1400
01390 PRINT"ASSOCIATED ONE-TAILED P = 1.00"
01400 END
```

PROGRAM G. ORTHOGONAL CONTRAST WEIGHTS

```
00010 EN=0
00020 EM=0
00030 B5=0
00040 B6=0
00050 B7=0
00060 PRINT"COMPUTATION OF ORTHOGONAL COEFFICIENTS (LINEAR AND QUADRATIC
)"
00070 PRINT"ENTER NUMBER OF LEVELS OF ATTRIBUTE"
00080 INPUT L
00090 DIM V (450), N (450), V2 (450), B1 (450), B3 (450), B4 (450), L (4
50)
00100 DIM Q (450)
00110 FOR X=1 TO L
00120 PRINT"ENTER NUMERICAL VALUE OF ATTRIBUTE LEVEL ";X
00130 INPUT V(X)
00140 PRINT"ENTER NUMBER OF EXEMPLARS OF ATTRIBUTE LEVEL ";X
00150 INPUT N(X)
00160 EN=EN+N(X)
00170 EM=EM+N(X)*V(X)
00180 V2(X)=V(X)**2
00190 NEXT X
00200 A=-EM/EN
00210 FOR X=1 TO L
```

```
00220 L(X)=(A+V(X))*EN
00230 B1(X)=N(X)*V(X)*L(X)
00240 B3(X)=N(X)*L(X)*V2(X)
00250 B4(X)=N(X)*V2(X)
00260 B5=B5+B1(X)
00270 B6=B6+B3(X)
00280 B7=B7+B4(X)
00290 NEXT X
00300 B2=-B6/B5
00310 A2=(-EM*B2-B7)/EN
00320 FOR X=1 TO L
00330 Q(X)=A2+B2*V(X)+V2(X)
00340 NEXT X
00350 PRINT"ORTHOGONAL COEFFICIENTS"
00360 PRINT"USER MAY SIMPLIFY COEFFICIENTS BY SOME COMMON DENOMINATOR"
00370 PRINT"(E.G., DIVIDE EVEN COEFFICIENTS BY 2)"
00380 FOR X=1 TO L
00390 PRINT"LINEAR COEFFICIENT FOR LEVEL ";X;"    = ";L(X)
00400 PRINT"QUADRATIC COEFFICIENT FOR LEVEL ";X;" = ";Q(X)
00410 PRINT"ENTER TO CONTINUE"
00420 INPUT A$
00430 NEXT X
00440 END
```

PROGRAM H. COMPARISON (FOCUSED) OF SIGNIFICANCE LEVELS (THREE OR MORE STUDIES)

```
00010  PRINT" META-ANALYSIS COMPARING SIGNIFICANCE LEVELS"
00020 PRINT" (FOCUSED TEST)"
00030 PRINT" (THREE OR MORE STUDIES)"
00040 DE$="ENTER DIRECTION OF EFFECT ('+' IF EXPECTED, '-' IF UNEXPECTED
)"
00050 N=0
00060 ZN=0
00070 W2=0
00080 PRINT"ENTER THE NUMBER OF STUDIES TO BE INCLUDED"
00090 INPUT G
00100 DIM ZM (500), W (500)
00110 N=N+1
00120 PRINT"ENTER TYPE OF STATISTIC FOR DATA SET #" ;N
00130 PRINT "(T=T-TEST; F=F-TEST; X=CHI-SQUARE; R=CORR COEF; P=PROB VALU
E)"
00140 INPUT A$
00150 IF A$="T" THEN GOTO 220
00160 IF A$="F" THEN GOTO 310
00170 IF A$="X" THEN GOTO 390
00180 IF A$="R" THEN GOTO 460
00190 IF A$="P" THEN GOTO 540
00200 N=N-1
00210 GOTO 110
```

```
00220 PRINT"ENTER T-TEST FOR DATA SET #";N
00230 INPUT S
00240 PRINT"ENTER DF ASSOCIATED WITH THIS T-TEST"
00250 INPUT DF
00260 PRINT DE$
00270 INPUT E$
00280 Z=(DF*LOG((1+(S**2/DF))))**.5*(1-(1/(2*DF)))**.5
00290 IF E$="-" THEN Z=-Z
00300 GOTO 630
00310 PRINT"ENTER F-TEST FOR DATA SET #";N
00320 INPUT S
00330 PRINT"ENTER DF(ERROR) ASSOCIATED WITH THIS F-TEST"
00340 INPUT DF
00350 PRINT DE$
00360 INPUT E$
00370 S=SQR(S)
00380 GOTO 280
00390 PRINT"ENTER CHI-SQUARE FOR DATA SET #";N
00400 INPUT S
00410 PRINT DE$
00420 INPUT E$
00430 Z=SQR(S)
00440 IF E$="-" THEN Z=-Z
00450 GOTO 630
00460 PRINT"ENTER CORRELATION COEFFICIENT FOR DATA SET #";N
00470 INPUT S
```

(Continued)

PROGRAM H *(Continued)*

```
00480 PRINT"ENTER DF ASSOCIATED WITH THIS CORRELATION"
00490 INPUT DF
00500 PRINT DE$
00510 INPUT E$
00520 S=(S*SQR(DF))/SQR(1-S**2)
00530 GOTO 280
00540 PRINT"ENTER ONE-TAILED P VALUE FOR DATA SET #";N
00550 INPUT P
00560 IF P >.5 THEN X=1
00570 IF P > .5 THEN P=1-P
00580 Q=SQR(LOG(1/(P**2)))
00590 Z1=(2.515517+.802853*Q+.010328*(Q**2))
00600 Z2=(1+1.432788*Q+.189269*(Q**2)+.001308*(Q**3))
00610 Z=Q-(Z1/Z2)
00620 IF X=1 THEN Z=-Z
00630 PRINT"ENTER CONTRAST WEIGHT"
00640 INPUT W
00650 PRINT"Z = ";Z
00660 ZM(N)=Z
00670 W(N)=W
00680 W2=W2+W**2
00690 Z=0
00700 W=0
00710 X=0
00720 E$=""
```

```
00730 IF N < G THEN GOTO 110
00740 FOR X=1 TO N
00750 ZN=ZN+ZM(X)*W(X)
00760 NEXT X
00770 ZM=ABS(ZN/SQR(W2))
00780 PRINT"* * * META-ANALYTIC STATISTICS"
00790 PRINT"Z FOR COMPARISON = ";ZM
00800 P1=(1+.049867347*ZM+.0211410061*ZM**2+.0032776263*ZM**3)
00810 P2=(.0000380036*ZM**4+.0000488906*ZM**5+.000005383*ZM**6)
00820 P=.5*(P1+P2)**-16
00830 IF ZM < 6.03788 THEN PRINT "ASSOCIATED ONE-TAILED P= ";P
00840 IF ZM < 6.03788 THEN GOTO 860
00850 PRINT"ASSOCIATED ONE-TAILED P < .000000001"
00860 END
```

PROGRAM I. COMPARISON (FOCUSED) OF EFFECT SIZES (THREE OR MORE STUDIES)

```
00010 PRINT " META-ANALYSIS COMPARING EFFECT SIZES"
00020 PRINT" (FOCUSED TEST)"
00030 PRINT" (THREE OR MORE STUDIES)"
00040 DE$="ENTER DIRECTION OF EFFECT ('+' IF EXPECTED, '-' IF UNEXPECTED
)"
00050 N=0
00060 RN=0
00070 RD=0
00080 PRINT "ENTER THE NUMBER OF STUDIES TO BE INCLUDED"
00090 INPUT G
00100 DIM AR (500), W (500), NR (500)
00110 N=N+1
00120 E$=""
00130 PRINT "DATA SET #";N
00140 PRINT "ENTER TYPE OF STATISTIC FOR DATA SET # ";N
00150 PRINT "(T=T-TEST; F=F-TEST; X=CHI-SQUARE; R=CORR COEF; P=PROB VALU
E)"
00160 INPUT A$
00170 IF A$="T" THEN GOTO 230
00180 IF A$="F" THEN GOTO 340
00190 IF A$="X" THEN GOTO 450
00200 IF A$="R" THEN GOTO 720
00210 IF A$="P" THEN GOTO 570
```

```
00220 GOTO 130
00230 PRINT "ENTER T-TEST FOR DATA SET#";N
00240 INPUT S
00250 PRINT "ENTER DF ASSOCIATED WITH THIS T-TEST"
00260 INPUT DF
00270 PRINT DE$
00280 INPUT E$
00290 R=SQR(S**2/(S**2+DF))
00300 IF E$="-" THEN R=-R
00310 D=SQR((4*R**2)/(1-R**2))
00320 IF E$="-" THEN D=-D
00330 GOTO 820
00340 PRINT "ENTER F-TEST FOR DATA SET #";N
00350 INPUT S
00360 PRINT "ENTER DF(ERROR) ASSOCIATED WITH THIS F-TEST"
00370 INPUT DF
00380 PRINT DE$
00390 INPUT E$
00400 R=SQR(S/(S+DF))
00410 IF E$="-" THEN R=-R
00420 D=SQR((4*R**2)/(1-R**2))
00430 IF E$="-" THEN D=-D
00440 GOTO 820
00450 PRINT "ENTER CHI-SQUARE FOR DATA SET #";N
00460 INPUT S
00470 PRINT "ENTER N ASSOCIATED WITH THIS CHI-SQUARE"
```

(Continued)

```
00480 INPUT DF
00490 PRINT DE$
00500 INPUT E$
00510 R=SQR(S/DF)
00520 IF E$="-" THEN R=-R
00530 D=SQR((4*R**2)/(1-R**2))
00540 IF E$="-" THEN D=-D
00550 DF=DF-2
00560 GOTO 820
00570 PRINT "ENTER ONE-TAILED P VALUE FOR DATA SET # ";N
00580 INPUT S
00590 PRINT "ENTER N ASSOCIATED WITH THIS PROBABILITY VALUE"
00600 INPUT DF
00610 PRINT DE$
00620 INPUT E$
00630 IF S >.5 THEN S=1-S
00640 Q=SQR(LOG(1/(S**2)))
00650 Z1=(2.515517+.802853*Q+.010328*(Q**2))
00660 Z2=(1+1.432788*Q+.189269*(Q**2)+.001308*(Q**3))
00670 Z=Q-(Z1/Z2)
00680 R=SQR(Z**2/DF)
00690 IF E$="-" THEN R=-(ABS(R))
00700 DF=DF-2
00710 GOTO 800
00720 PRINT "ENTER CORRELATION COEFFICIENT FOR DATA SET #";N
```

```
00730 INPUT S
00740 PRINT "ENTER DF ASSOCIATED WITH THIS CORRELATION COEFFICIENT"
00750 INPUT DF
00760 PRINT DE$
00770 INPUT E$
00780 R=S
00790 IF E$="-" THEN R=-(ABS(R))
00800 D=SQR((4*R**2)/(1-R**2))
00810 IF E$="-" THEN D=-D
00820 PRINT"ENTER CONTRAST WEIGHT"
00830 INPUT W
00840 ZR=((LOG(1+R))-(LOG(1-R)))/2
00850 AR(N)=ZR
00860 NR(N)=DF-1
00870 W(N)=W
00880 B=.5+(R/2)
00890 PRINT "EFFECT SIZE STATISTICS FOR DATA SET #";N
00900 PRINT "CORRELATION COEFFICIENT:          R = ";R
00910 PRINT "PROPORTION OF VARIANCE ACCOUNTED FOR:  R2 = ";R**2
00920 PRINT "BINOMIAL EFFECT SIZE DISPLAY:    FROM ";1-B;"TO ";B
00930 PRINT "STANDARDIZED DIFFERENCE BETWEEN MEANS:   D = ";D
00940 X=0
00950 R=0
00960 ZR=0
00970 D=0
```

(Continued)

PROGRAM I *(Continued)*

```
00980 B=0
00990 W=0
01000 PRINT "ENTER TO CONTINUE"
01010 INPUT A$
01020 IF N < G  THEN GOTO 110
01030 FOR X=1 TO N
01040 RN=RN+(AR(X)*W(X))
01050 RD=RD+(W(X)**2/NR(X))
01060 NEXT X
01070 RZ=ABS(RN/SQR(RD))
01080 PRINT " * * * META-ANALYTIC STATISTICS"
01090 PRINT"Z FOR COMPARISON = ";RZ
01100 Z=RZ
01110 P1=(1+.049867347*Z+.0211410061*Z**2+.0032776263*Z**3)
01120 P2=(.0000380036*Z**4+.0000488906*Z**5+.000005383*Z**6)
01130 P=.5*(P1+P2)**-16
01140 IF Z < 6.03788 THEN PRINT"ASSOCIATED ONE-TAILED P = ";P
01150 IF Z < 6.03788 THEN GOTO 1170
01160 PRINT"ASSOCIATED ONE-TAILED P < .000000001"
01170 END
```

PROGRAM J. COMBINATION OF SIGNIFICANCE LEVELS (THREE OR MORE STUDIES)

```
00010  PRINT" META-ANALYSIS COMBINING SIGNIFICANCE LEVELS"
00020 PRINT" (THREE OR MORE STUDIES)"
00030 DE$="ENTER DIRECTION OF EFFECT ('+' IF EXPECTED, '-' IF UNEXPECTED
)"
00040 N=0
00050 ZM=0
00060 ZD=0
00070 ZF=0
00080 PRINT"ENTER THE NUMBER OF STUDIES TO BE INCLUDED"
00090 INPUT G
00100 DIM ZM (500)
00110 N=N+1
00120 PRINT"ENTER TYPE OF STATISTIC FOR DATA SET #" ;N
00130 PRINT "(T=T-TEST; F=F-TEST; X=CHI-SQUARE; R=CORR COEF; P=PROB VALU
E)"
00140 INPUT A$
00150 IF A$="T" THEN GOTO 220
00160 IF A$="F" THEN GOTO 310
00170 IF A$="X" THEN GOTO 390
00180 IF A$="R" THEN GOTO 460
00190 IF A$="P" THEN GOTO 540
00200 N=N-1
00210 GOTO 110
```

(Continued)

PROGRAM J *(Continued)*

```
00220 PRINT"ENTER T-TEST FOR DATA SET #";N
00230 INPUT S
00240 PRINT"ENTER DF ASSOCIATED WITH THIS T-TEST"
00250 INPUT DF
00260 PRINT DE$
00270 INPUT E$
00280 Z=(DF*LOG((1+(S**2/DF))))**.5*(1-(1/(2*DF)))**.5
00290 IF E$="-" THEN Z=-Z
00300 GOTO 630
00310 PRINT"ENTER F-TEST FOR DATA SET #";N
00320 INPUT S
00330 PRINT"ENTER DF(ERROR) ASSOCIATED WITH THIS F-TEST"
00340 INPUT DF
00350 PRINT DE$
00360 INPUT E$
00370 S=SQR(S)
00380 GOTO 280
00390 PRINT"ENTER CHI-SQUARE FOR DATA SET #";N
00400 INPUT S
00410 PRINT DE$
00420 INPUT E$
00430 Z=SQR(S)
00440 IF E$="-" THEN Z=-Z
00450 GOTO 630
00460 PRINT"ENTER CORRELATION COEFFICIENT FOR DATA SET #";N
00470 INPUT S
```

```
00480 PRINT"ENTER DF ASSOCIATED WITH THIS CORRELATION"
00490 INPUT DF
00500 PRINT DE$
00510 INPUT E$
00520 S=(S*SQR(DF))/SQR(1-S**2)
00530 GOTO 280
00540 PRINT"ENTER ONE-TAILED P VALUE FOR DATA SET #";N
00550 INPUT P
00560 IF P >.5 THEN X=1
00570 IF P > .5 THEN P=1-P
00580 Q=SQR(LOG(1/(P**2)))
00590 Z1=(2.515517+.802853*Q+.010328*(Q**2))
00600 Z2=(1+1.432788*Q+.189269*(Q**2)+.001308*(Q**3))
00610 Z=Q-(Z1/Z2)
00620 IF X=1 THEN Z=-Z
00630 PRINT"ENTER WEIGHT"
00640 INPUT W
00650 PRINT"Z = ";Z
00660 ZM=ZM+Z*W
00670 ZF=ZF+Z
00680 ZD=ZD+W**2
00690 Z=0
00700 W=0
00710 X=0
00720 E$=""
00730 IF N < G THEN GOTO 110
```

(Continued)

PROGRAM J *(Continued)*

```
00740 ZM=ABS(ZM/SQR(ZD))
00750 FS=(ZF/1.645)**2-N
00760 PRINT"* * * META-ANALYTIC STATISTICS"
00770 PRINT"Z FOR COMBINATION = ";ZM
00780 Z1=(1+.049867347*ZM+.0211410061*(ZM**2)+.0032776263*(ZM**3))
00790 Z2=(.0000380036*(ZM**4)+.0000488906*(ZM**5)+.000005383*(ZM**6))
00800 P=.5*(Z1+Z2)**-16
00810 IF ZM < 6.03788 THEN PRINT"ASSOCIATED ONE-TAILED P = ";P
00820 IF ZM >= 6.03788 THEN PRINT"ASSOCIATED ONE-TAILED P<.000000001"
00830 PRINT"FAIL SAFE NUMBER (P = .05) = ";FS
00840 END
```

PROGRAM K. COMBINATION OF EFFECT SIZES (THREE OR MORE STUDIES)

```
00010 PRINT "    META-ANALYSIS COMBINING EFFECT SIZES"
00020 PRINT "    (THREE OR MORE STUDIES)"
00030 DE$="ENTER DIRECTION OF EFFECT ('+' IF EXPECTED, '-' IF UNEXPECTED
)"
00040 N=0
00050 ZN=0
00060 W2=0
00070 PRINT "ENTER THE NUMBER OF STUDIES TO BE INCLUDED"
00080 INPUT G
00090 DIM ZM (500)
00100 N=N+1
00110 E$=""
00120 PRINT "DATA SET #";N
00130 PRINT "ENTER TYPE OF STATISTIC FOR DATA SET # ";N
00140 PRINT "(T=T-TEST; F=F-TEST; X=CHI-SQUARE; R=CORR COEF; P=PROB VALU
E)"
00150 INPUT A$
00160 IF A$="T" THEN GOTO 220
00170 IF A$="F" THEN GOTO 330
00180 IF A$="X" THEN GOTO 440
00190 IF A$="R" THEN GOTO 710
00200 IF A$="P" THEN GOTO 560
00210 GOTO 120
```

(Continued)

PROGRAM K *(Continued)*

```
00220 PRINT "ENTER T-TEST FOR DATA SET#";N
00230 INPUT S
00240 PRINT "ENTER DF ASSOCIATED WITH THIS T-TEST"
00250 INPUT DF
00260 PRINT DE$
00270 INPUT E$
00280 R=SQR(S**2/(S**2+DF))
00290 IF E$="-" THEN R=-R
00300 D=SQR((4*R**2)/(1-R**2))
00310 IF E$="-" THEN D=-D
00320 GOTO 810
00330 PRINT "ENTER F-TEST FOR DATA SET #";N
00340 INPUT S
00350 PRINT "ENTER DF(ERROR) ASSOCIATED WITH THIS F-TEST"
00360 INPUT DF
00370 PRINT DE$
00380 INPUT E$
00390 R=SQR(S/(S+DF))
00400 IF E$="-" THEN R=-R
00410 D=SQR((4*R**2)/(1-R**2))
00420 IF E$="-" THEN D=-D
00430 GOTO 810
00440 PRINT "ENTER CHI-SQUARE FOR DATA SET #";N
00450 INPUT S
00460 PRINT "ENTER N ASSOCIATED WITH THIS CHI-SQUARE"
00470 INPUT DF
```

```
00480 PRINT DE$
00490 INPUT E$
00500 R=SQR(S/DF)
00510 IF E$="-" THEN R=-R
00520 D=SQR((4*R**2)/(1-R**2))
00530 IF E$="-" THEN D=-D
00540 DF=DF-2
00550 GOTO 810
00560 PRINT "ENTER ONE-TAILED P VALUE FOR DATA SET # ";N
00570 INPUT S
00580 PRINT "ENTER N ASSOCIATED WITH THIS PROBABILITY VALUE"
00590 INPUT DF
00600 PRINT DE$
00610 INPUT E$
00620 IF S >.5 THEN S=1-S
00630 Q=SQR(LOG(1/(S**2)))
00640 Z1=(2.515517+.802853*Q+.010328*(Q**2))
00650 Z2=(1+1.432788*Q+.189269*(Q**2)+.001308*(Q**3))
00660 Z=Q-(Z1/Z2)
00670 R=SQR(Z**2/DF)
00680 IF E$="-" THEN R=-(ABS(R))
00690 DF=DF-2
00700 GOTO 790
00710 PRINT "ENTER CORRELATION COEFFICIENT FOR DATA SET #";N
00720 INPUT S
00730 PRINT "ENTER DF ASSOCIATED WITH THIS CORRELATION COEFFICIENT"
```

(Continued)

PROGRAM K *(Continued)*

```
00740 INPUT DF
00750 PRINT DE$
00760 INPUT E$
00770 R=S
00780 IF E$="-" THEN R=-(ABS(R))
00790 D=SQR((4*R**2)/(1-R**2))
00800 IF E$="-" THEN D=-D
00810 PRINT "ENTER WEIGHT"
00820 INPUT W1
00830 ZR=((LOG(1+R))-(LOG(1-R)))/2
00840 ZN =ZN+ZR*W1
00850 W2=W2+W1
00860 B=.5+(R/2)
00870 PRINT "EFFECT SIZE STATISTICS FOR DATA SET #";N
00880 PRINT "CORRELATION COEFFICIENT:          R = ";R
00890 PRINT "PROPORTION OF VARIANCE ACCOUNTED FOR:  R2 = ";R**2
00900 PRINT "BINOMIAL EFFECT SIZE DISPLAY:    FROM ";1-B;"TO ";B
00910 PRINT "STANDARDIZED DIFFERENCE BETWEEN MEANS:   D = ";D
00920 X=0
00930 R=0
00940 Z=0
00950 W1=0
00960 D=0
00970 PRINT "ENTER TO CONTINUE"
```

```
00980 INPUT A$
00990 IF N < G  THEN GOTO 100
01000 PRINT " * * * META-ANALYTIC STATISTICS"
01010 ZN=ZN/W2
01020 RN=((EXP(ZN))**2-1)/((EXP(ZN))**2+1)
01030 B=.5+(RN/2)
01040 D=SQR((4*RN**2)/(1-RN**2))
01050 PRINT "FISHER Z FOR COMBINATION:            Z = ";ZN
01060 PRINT "MEAN CORRELATION COEFFICIENT:            R = ";RN
01070 PRINT "PROPORTION OF VARIANCE ACCOUNTED FOR:  R2 = ";RN**2
01080 PRINT "BINOMIAL EFFECT SIZE DISPLAY:    FROM";1-B;" TO ";B
01090 PRINT "MEAN STANDARDIZED DIFFERENCE BETWEEN MEANS:  D = ";D
01100 END
```

PROGRAM L. CORRELATION MATRIX, PARTIAL CORRELATION AND MULTIPLE CORRELATION

```
00010 PRINT"CORRELATION, PARTIAL CORR. AND MULTIPLE CORR."
00020 PRINT"ENTER NUMBER OF STUDIES (ROWS)"
00030 INPUT R
00040 PRINT"ENTER NUMBER OF VARIABLES (COLUMNS)"
00050 INPUT C
00060 DIM DA (100,8), V1 (1,100), V2 (100,1), MV (1,8)
00070 DIM MX (100,8), ID (8,8), SV (1,8), Z (100,8)
00080 DIM ZT (8,100), V$ (15), W$ (3), P (3), M(3)
00090 FOR X = 1 TO C
00100 PRINT"ENTER NAME OF VARIABLE #";X
00110 INPUT V$ (X)
00120 NEXT X
00130 FOR I = 1 TO R
00140 PRINT"INPUT DATA FOR STUDY #";I
00150 FOR J = 1 TO C
00160 INPUT DA (I,J)
00170 NEXT J
00180 NEXT I
00190 I = 1
00200 FOR J = 1 TO R
00210 V1 (I,J) = 1
00220 NEXT J
00230 FOR I = 1 TO R
```

```
00240 J=1
00250 V2 (I,J) = 1
00260 NEXT I
00270 I=1
00280 FOR J = 1 TO C
00290 MV (I,J) = 0
00300 FOR K = 1 TO R
00310 MV (I,J) = MV (I,J)+V1 (I,K)*DA(K,J)*(1/R)
00320 NEXT K
00330 NEXT J
00340 FOR J = 1 TO C
00350 I = 1
00360 PRINT"MEAN ";V$(J);" = ";MV(I,J),
00370 NEXT J
00380 PRINT"ENTER TO CONTINUE"
00390 INPUT A$
00400 FOR I = 1 TO R
00410 FOR J = 1 TO C
00420 MX (I,J) = 0
00430 K = 1
00440 MX(I,J)=MX(I,J)-(V2(I,K)*MV(K,J))
00450 NEXT J
00460 NEXT I
00470 FOR J= 1 TO C
00480 FOR I = 1 TO R
00490 Z(I,J)=MX(I,J)+DA(I,J)
```

(Continued)

PROGRAM L *(Continued)*

```
00500 NEXT I
00510 NEXT J
00520 FOR J = 1 TO C
00530 FOR I = 1 TO C
00540 IF I = J THEN ID(I,J)=1 ELSE ID(I,J)=0
00550 NEXT I
00560 NEXT J
00570 FOR J = 1 TO C
00580 FOR I = 1 TO R
00590 MX(I,J)=Z(I,J)*Z(I,J)
00600 NEXT I
00610 NEXT J
00620 I=1
00630 FOR J = 1 TO C
00640 MV (I,J) = 0
00650 FOR K = 1 TO R
00660 MV(I,J)=MV(I,J)+V1(I,K)*MX(K,J)
00670 NEXT K
00680 NEXT J
00690 I=1
00700 FOR J = 1 TO C
00710 SV(I,J)=SQR(MV(I,J)/(R-1))
00720 NEXT J
00730 I=1
00740 FOR J = 1 TO C
```

```
00750 PRINT"S.D. ";V$(J);" = ";SV(I,J),
00760 NEXT J
00770 PRINT"ENTER TO CONTINUE"
00780 INPUT A$
00790 FOR I = 1 TO C
00800 FOR J = 1 TO C
00810 K = 1
00820 IF I = J THEN ID(I,J)=SV(K,J) ELSE  ID(I,J)=0
00830 NEXT J
00840 NEXT I
00850 FOR I = 1 TO C
00860 FOR J = 1 TO C
00870 IF I = J THEN ID(I,J)=1/ID(I,J) ELSE ID(I,J)=0
00880 NEXT J
00890 NEXT I
00900 FOR I = 1 TO R
00910 FOR J = 1 TO C
00920 MX(I,J)=Z(I,J)
00930 NEXT J
00940 NEXT I
00950 FOR I = 1 TO R
00960 FOR J = 1 TO C
00970 Z(I,J)=0
00980 FOR K = 1 TO C
00990 Z(I,J)=Z(I,J)+MX(I,K)*ID(K,J)
01000 NEXT K
```

(Continued)

PROGRAM L *(Continued)*

```
01010 NEXT J
01020 NEXT I
01030 FOR I = 1 TO R
01040 FOR J = 1 TO C
01050 ZT(J,I)=Z(I,J)
01060 NEXT J
01070 NEXT I
01080 FOR I = 1 TO C
01090 FOR J = 1 TO C
01100 ID(I,J)=0
01110 FOR K = 1 TO R
01120 ID(I,J)=ID(I,J)+ZT(I,K)*Z(K,J)
01130 NEXT K
01140 NEXT J
01150 NEXT I
01160 FOR J = 1 TO C
01170 FOR I = 1 TO C
01180 ID(I,J)=ID(I,J)*(1/(R-1))
01190 NEXT I
01200 NEXT J
01210 PRINT"PRODUCT MOMENT CORRELATION MATRIX"
01220 FOR I = 1 TO C
01230 FOR J = 1 TO C
01240 PRINT" ";V$(I);":";V$(J);"     ";ID(I,J)
01250 NEXT J
```

```
01260 PRINT"ENTER TO CONTINUE"
01270 INPUT A$
01280 NEXT I
01290 PRINT"ENTER 'P' TO COMPUTE PARTIAL CORRELATION"
01300 PRINT"ENTER 'M' TO COMPUTE MULTIPLE CORRELATION"
01310 PRINT"ENTER 'R' TO DISPLAY CORRELATION MATRIX"
01320 PRINT"ENTER 'E' TO END"
01330 INPUT B$
01340 IF B$="P" THEN GOTO 1390
01350 IF B$="M" THEN GOTO 1580
01360 IF B$="R" THEN GOTO 1210
01370 IF B$="E" THEN GOTO 1770
01380 GOTO 1290
01390 PRINT"PARTIAL CORRELATION"
01400 PRINT"ENTER THE NAME OF THE FIRST VARIABLE"
01410 INPUT W$(1)
01420 PRINT"ENTER THE NAME OF THE SECOND VARIABLE"
01430 INPUT W$(2)
01440 PRINT"ENTER THE NAME OF THE VARIABLE TO BE PARTIALLED OUT"
01450 INPUT W$(3)
01460 FOR X=1 TO 3
01470 FOR Y = 1 TO C
01480 IF V$(Y)=W$(X) THEN P(X)=Y
01490 NEXT Y
01500 NEXT X
01510 M(1)=ID(P(1),P(2))
```

(Continued)

PROGRAM L *(Continued)*

```
01520 M(2)=ID(P(2),P(3))
01530 M(3)=ID(P(1),P(3))
01540 P=(M(1)-M(3)*M(2))/SQR((1-M(3)**2)*(1-M(2)**2))
01550 PRINT"PARTIAL CORRELATION COEFFICIENT"
01560 PRINT"R (";W$(1);",";W$(2);".";W$(3);") = ";P
01570 GOTO 1290
01580 PRINT"MULTIPLE CORRELATION"
01590 PRINT"ENTER THE NAME OF THE 'CRITERION' VARIABLE"
01600 INPUT W$(1)
01610 PRINT"ENTER THE NAME OF THE FIRST 'PREDICTOR' VARIABLE"
01620 INPUT W$(2)
01630 PRINT"ENTER THE NAME OF THE SECOND 'PREDICTOR' VARIABLE"
01640 INPUT W$(3)
01650 FOR X = 1 TO 3
01660 FOR Y = 1 TO C
01670 IF V$(Y)=W$(X) THEN P(X)=Y
01680 NEXT Y
01690 NEXT X
01700 M(1)=ID(P(1),P(2))
01710 M(2)=ID(P(2),P(3))
01720 M(3)=ID(P(1),P(3))
01730 P=SQR((M(1)**2+M(3)**2-(2*M(1)*M(3)*M(2)))/(1-M(2)**2))
01740 PRINT"MULTIPLE CORRELATION COEFFICIENT"
01750 PRINT"R (";W$(1);".";W$(2);",";W$(3);") = ";P
01760 GOTO 1290
01770 END
```

PROGRAM M. DISJOINT CLUSTER ANALYSIS

```
00010 PRINT"DISJOINT CLUSTER ANALYSIS OF EFFECT SIZES (R)"
00020 PRINT"ENTER NUMBER OF STUDIES TO BE INCLUDED"
00030 INPUT NS
00040 DIM O (300), S (300), U (300), T (300), G (301)
00050 A=0
00060 Y=0
00070 FOR X = 1 TO NS
00080 PRINT"ENTER EFFECT SIZE (R) FOR DATA SET #";X
00090 INPUT O (X)
00100 PRINT"ENTER SAMPLE SIZE FOR DATA SET #";X
00110 INPUT S (X)
00120 T(X)=X
00130 NEXT X
00140 FOR X = 1 TO NS
00150 A=A+SQR(S(X)-3)
00160 NEXT X
00170 A=(A/NS)**2
00180 FOR X = 1 TO NS
00190 U(X)=SQR(A)*((LOG(1+O(X)))-(LOG(1-O(X))))/2
00200 NEXT X
00210 FOR X = 1 TO (NS-1)
00220 WQ=X+1
00230 M=X
00240 FOR J = WQ TO NS
```

(Continued)

```
00250 IF U(M)=>U(J) THEN GOTO 270
00260 M=J
00270 NEXT J
00280 T=U(X)
00290 S=T(X)
00300 R=O(X)
00310 U(X)=U(M)
00320 T(X)=T(M)
00330 O(X)=O(M)
00340 U(M)=T
00350 T(M)=S
00360 O(M)=R
00370 NEXT X
00380 FOR X = 1 TO (NS-1)
00390 WQ=X+1
00400 G(X)=U(X)-U(WQ)
00410 NEXT X
00420 PRINT"ENTER SIGNIFICANCE LEVEL ('.10','.05','.01', OR 'END')"
00430 INPUT SF$
00440 IF SF$ = ".10" THEN GOTO 650
00450 IF SF$ = ".05" THEN GOTO 680
00460 IF SF$ = ".01" THEN GOTO 710
00470 IF SF$="END" THEN GOTO 740
00480 GOTO 420
```

```
00490 W=1
00500 FOR X = 1 TO NS
00510 IF G(X)<MG THEN S(X)=W
00520 IF G(X)=>MG THEN S(X)=W
00530 IF G(X)=>MG THEN W=W+1
00540 NEXT X
00550 PRINT"LISTING OF CLUSTERS AT P = ";SF$;" LEVEL OF SIGNIFICANCE"
00560 Y=1
00570 PRINT"CLUSTER ";Y
00580 FOR X = 1 TO NS
00590 IF S(X)=Y THEN PRINT"    DATA SET # ";T(X);" (R = ";O(X);")"
00600 NEXT X
00610 PRINT"ENTER TO CONTINUE"
00620 INPUT A$
00630 IF Y<=W THEN Y=Y+1
00640 IF Y<=W THEN GOTO 570 ELSE GOTO 420
00650 M1=-.563637*LOG(NS)+.000140264*((LOG(NS))**2)+.00851203*((LOG(NS))
**3)
00660 MG=2.76508-(.00274963*NS)+M1
00670 GOTO 490
00680 M1=-1.1441*LOG(NS)+.216853*(LOG(NS))**2-.023789*(LOG(NS))**3
00690 MG=3.49313+.00470484*NS+M1
00700 GOTO 490
00710 M1=-1.06998*LOG(NS)+.0863124*(LOG(NS))**2+.00691388*(LOG(NS))**3
00720 MG=4.2229-.00331781*NS+M1
00730 GOTO 490
00740 END
```

PROGRAM N. MODEL ESTIMATION

```
00010 PRINT"       META-ANALYTIC MODEL ESTIMATION"
00020 PRINT"ENTER THE NUMBER OF COMPARISONS"
00030 INPUT U
00040 PRINT"ENTER THE NAME OF THE INDEPENDENT VARIABLE"
00050 INPUT N$
00060 DIM X (500), Y (500), S (250), L (250)
00070 N=0
00080 N=N+1
00090 PRINT"COMPARISON # ";N
00100 PRINT"ENTER NUMERICAL VALUE FOR SMALLER ";N$
00110 INPUT NS
00120 PRINT"ENTER NUMERICAL VALUE FOR LARGER ";N$
00130 INPUT NL
00140 PRINT"ENTER STUDY OUTCOME"
00150 INPUT SO
00160 S(N)=NS
00170 L(N)=NL
00180 Y(N)=SO
00190 IF N < U THEN GOTO 80
00200 PRINT"MODEL SPECIFICATION"
00210 PRINT"ENTER     1 = (LARGER ";N$;" / SMALLER ";N$;")"
00220 PRINT"          2 = LOG(LARGER ";N$;" / SMALLER ";N$;")"
00230 PRINT"          3 = (LARGER ";N$;" / SMALLER ";N$;") ** X"
00240 PRINT"                 (USER SUPPLIES THE EXPONENT 'X')"
```

```
00250 PRINT"              4 = END"
00260 INPUT C
00270 IF C=4 THEN GOTO 1040
00280 IF C=3 THEN PRINT"ENTER EXPONENT"
00290 IF C=3 THEN INPUT E
00300 FOR W=1 TO U
00310 IF C=1 THEN X(W)=L(W)/S(W)
00320 NEXT W
00330 FOR W=1 TO U
00340 IF C=2 THEN X(W)=LOG(L(W)/S(W))
00350 IF C=3 THEN X(W)=(L(W)/S(W))**E
00360 NEXT W
00370 EX=0
00380 EY=0
00390 XY=0
00400 X2=0
00410 Y2=0
00420 VX=0
00430 VY=0
00440 FOR  W=1 TO U
00450 EX=EX+X(W)
00460 EY=EY+Y(W)
00470 XY=XY+X(W)*Y(W)
00480 X2=X2+X(W)**2
00490 Y2=Y2+Y(W)**2
00500 NEXT W
```

(Continued)

```
00510 FOR I=1 TO U
00520 VX=VX+(X(I)-(EX/N))**2
00530 NEXT I
00540 SX=SQR(VX/N)
00550 FOR I=1 TO U
00560 VY=VY+(Y(I)-(EY/N))**2
00570 NEXT I
00580 SY=SQR(VY/N)
00590 B=((XY/N)-(EX/N)*(EY/N))/SX**2
00600 A=(EY/N)-B*(EX/N)
00610 R2=B**2*SX**2/SY**2
00620 R=SQR(R2)
00630 IF B<0 THEN R=-R
00640 PRINT"* * * TEST OF MODEL FIT"
00650 IF C=1 THEN PRINT"MODEL: (LARGER ";N$;" / SMALLER ";N$;")"
00660 IF C=2 THEN PRINT"MODEL: LOG(LARGER ";N$;" / SMALLER ";N$;")"
00670 IF C=3 THEN PRINT"MODEL: (LARGER ";N$;" / SMALLER ";N$;") ** ";E
00680 PRINT"MEAN PREDICTOR = ";EX/N
00690 PRINT"S.D. PREDICTOR = ";SX
00700 PRINT"MEAN STUDY OUTCOME = ";EY/N
00710 PRINT"S.D. STUDY OUTCOME = ";SY
00720 PRINT"COEFFICIENT OF REGRESSION (R) = ";R
00730 PRINT"COEFFICIENT OF DETERMINATION (R**2) = ";R2
00740 PRINT"FORMULA FOR  BEST-FITTING LINE:"
00750 PRINT"   Y = ";A;" + (";B;"X)"
00760 F=R2/((1-R2)/(N-2))
```

```
00770 PRINT"F-TEST FOR SIGNIFICANCE OF REGRESSION:"
00780 PRINT"    F(1,";N-2;") = ";F
00790 D2=N-2
00800 X=1
00810 IF F<1 THEN GOTO 860
00820 S=1
00830 T=D2
00840 Z=F
00850 GOTO 890
00860 S=D2
00870 T=1
00880 Z=1/F
00890 J=2/9/S
00900 K=2/9/T
00910 Y=ABS((1-K)*Z**(1/3)-1+J)/SQR(K*Z**(2/3)+J)
00920 IF T<4 THEN GOTO 960
00930 X=.5/(1+Y*(.196854+Y*(.115194+Y*(.000344+Y*.019527))))**4
00940 X=INT(X*10000+.5)/10000
00950 GOTO 980
00960 Y=Y*(1+.08*Y**4/T**3)
00970 GOTO 930
00980 IF F>=1 THEN GOTO 1000
00990 X=1-X
01000 PRINT"ASSOCIATED ONE-TAILED P = ";X
01010 PRINT"ENTER TO CONTINUE"
01020 INPUT A$
01030 GOTO 200
01040 END
```

6 Formulae Used in Meta-analysis Programs

The computational formulae for the Zs and the χ^2s for combination and comparison are presented in Chapter 1, and again in Chapter 4 along with numerical illustrations. Presented in this chapter are the formulae and algorithms which link the entered study data to these meta-analytic formulae.

The following procedure (Abramowitz & Stegun, 1964) is used to derive p from computed Z values:

$$p = .5\,(1 + .049867347Z + .0211410061Z^2 + .0032776263Z^3 + .0000380036Z^4 + .0000488906Z^5 + .000005383Z^6)^{-16}$$

The following procedure (Sokol & Rohlf, 1969) is used to derive p from computed χ^2 values:

$$df = \mathrm{k} - 1$$

$$KQ = \frac{\chi^{2(\mathrm{integer}(df+1)/2)}\,(\exp(-\chi^2/2))}{W}$$

where: W = the product of $(df)\,(df-2)\,(df-4)$...and so on until the multiplicand reaches or goes below "2"

if df is even then $J = 1$
if df is odd then $J = (2/\chi^2/3.1415927)$
$LQ = 1$, $M = 1$, $DX = df$
→ $DX = DX + 2$, $M = M(\chi^2)/DX$
if $M > .0000001$ then $LQ = LQ + M$ and repeat (return to →)
if $M < .0000001$ then $p = 1 - (J)\,(KQ)\,(LQ)$

For those programs dealing with significance levels, entered study data are converted directly into Zs. The following algorithm (Abramowitz & Stegun, 1964) is used to derive Z from an entered p value:

if $p > .5$ then $X = 1$ and $p = 1 - p$

$$Q = \sqrt{\log(1/p^2)}$$

$$Z = Q - \left(\frac{(2.515517 + .802853Q + .010328Q^2)}{(1 + 1.432788Q + .189269Q^2 + .001308Q^3)}\right)$$

if $X = 1$ then $Z = -Z$

Rosenthal (in press) summarizes a number of algorithms for converting other study statistics into Zs representing significance levels. The following algorithm (Wallace, 1959) is used to derive Z from an entered t:

$$Z = (df\ (\log_e(1 + (t^2/df\ ))))^{1/2}(1 - (1/(2df\ )))^{1/2}$$

The following algorithm is used to derive Z from an entered F:

$$Z = (df\ (\log_e(1 + (F/df\ ))))^{1/2}(1 - (1/(2df\ )))^{1/2}$$

The following algorithm is used to derive Z from an entered χ^2:

$$Z = \sqrt{\chi^2}$$

If the statistic entered for a study is an r, this r is first converted into a t using this algorithm:

$$t = \frac{r\ \sqrt{N-2}}{\sqrt{1-r^2}}$$

Then, this t is converted into a Z, using the algorithm presented above.

For those programs dealing with effect sizes, entered study data are converted into rs; these rs are then converted into r^2, $BESD$, d and the Fisher Z used in combination or comparison. Cohen (1965; 1977) provides a number of algorithms for converting study statistics into rs representing effect sizes. The following algorithm is used to derive r from an entered t:

$$r = \sqrt{\frac{t^2}{t^2 + df}}$$

The following algorithm is used to derive r from an entered F:

$$r = \frac{F}{F + df}$$

The following algorithm is used to derive r from an entered χ^2:

$$r = \sqrt{\frac{\chi^2}{N}}$$

If the statistic entered for a study is a p, this p is converted into a Z using the algorithm presented above. Then, the following algorithm is used to derive r from this Z:

$$r = \sqrt{\frac{Z^2}{N}}$$

Of course, if the statistic entered for a study is already an r, no such conversions are necessary. Once an r has been obtained for a study, r^2 and *BESD* (defined as a change from $(.50 - r/2)$ to $(.50 + r/2)$) are directly obtained. The following algorithm (Rosenthal, 1983) is used to derive d from r:

$$d = \sqrt{\frac{4r^2}{1-r^2}}$$

Fisher's transformation is used to derive a Fisher's Z from the obtained r:

$$Z = \tfrac{1}{2} \log_e \left(\frac{1+r}{1-r} \right)$$

The disjoint cluster analysis conducted by Program M uses linear approximations of the critical values for differences between rank-ordered effect sizes provided by Hedges and Olkin (1983, Table 1). A separate algorithm is used for each of three possible significance levels which can be chosen in Program M. The algorithm for the critical value for the significance level of $p = .10$ is:

$$\text{critical value}_{(p=.10)} = 2.76508 - .00274963k - .563637(\log(k)) + .000146264(\log(k))^2 + .00851203(\log(k))^3$$

The algorithm for the critical value for the significance level of $p = .05$ is:

$$\text{critical value}_{(p=.05)} = 3.49313 + .00470484k - 1.1441(\log(k)) + .216853(\log(k))^2 - .023789(\log(k))^3$$

The algorithm for the critical value for the significance level of $p = .01$ is:

$$\text{critical value}_{(p=.01)} = 4.2229 - .00331781k - 1.06998(\log(k)) + .0863124(\log(k))^2 + .00691388(\log(k))^3$$

The model estimation procedure of Program N involves regression analysis. Discussion of the computational formulae for regression analysis can be found in Kerlinger and Pedhazur (1973).

References

Abramowitz, M., & Stegun, I. A. (Eds.), *Handbook of Mathematical Functions*. Washington: National Bureau of Standards, 1964.

Atkinson, S. D., & Dolan, D. R. In search of research studies in on-line data bases. *Online*, 1983, *7*, 51–63.

Bakan, D. *On Method*. San Francisco: Jossey-Bass, 1967.

Bracey, G. W. Computers in Education: What the research shows. *Electronic Learning*, 1982, *2*, 51–54.

Campbell, D., & Stanley, J. *Experimental and Quasi-Experimental Designs for Research*. Chicago: Rand-McNally, 1963.

Chereb, D. M. Multiple regression analysis simplified. *Creative Computing*, 1979 (February), 126–129.

Cohen, J. Some statistical issues in psychological research. In B. B. Wolman (Ed.), *Handbook of Clinical Psychology*. New York: McGraw-Hill, 1965.

Cohen, J. *Statistical Power Analysis for the Behavioral Sciences*. New York: Academic Press, 1977.

Cohen, P. A. Validity of student ratings in psychology courses: A research synthesis. *Teaching of Psychology*, 1982, *9*, 78–82.

Cohen, P. A., Kulik, J. A., & Kulik, C. Education outcomes of tutoring: A meta-analysis of findings. *American Educational Research Journal*, 1982, *14*, 237–248.

Cook, T. D. & Leviton, L. C. Reviewing the literature: A comparison of traditional methods with meta-analysis. *Journal of Personality*, 1980, *48*, 449–472.

Cooper, H. M. Statistically combining independent studies: A meta-analysis of sex differences in conformity research. *Journal of Personality and Social Psychology*, 1979, *37*, 131–135.

Cooper, H. M. Scientific guidelines for conducting integrative research review. *Review of Educational Research*, 1982, *52*, 291–302.

Cooper, H. M. *The Integrative Research Review: A Social Science Approach*. Beverly Hills: Sage, in press.

Cooper, H. M., & Rosenthal, R. Statistical versus traditional procedures for summarizing research findings. *Psychological Bulletin*, 1980, *87*, 442–449.

DePaulo, B. M., Zuckerman, M., & Rosenthal, R. Detecting deception; Modality effects. In L. Wheeler (Ed.) *Review of Personality and Social Psychology.* New York: Sage, 1980.

Eagly, A. H., & Carli, L. L. Sex of researchers and sex-typed communications as determinants of sex differences in influenceability: A meta-analysis of social influence studies. *Psychological Bulletin,* 1981, *90,* 1–20.

Eysenck, J. J. An exercise in mega-silliness. *American Psychologist,* 1978, *33,* 517.

Federighi, E. T. Extended tables of the percentage points of student's *t*-distribution. *Journal of the American Statistical Association,* 1959, *54,* 683–688.

Fiske, D. W. The meta-analytic revolution in outcome research. *Journal of Consulting and Clinical Psychology,* 1983, *51,* 65–70.

Gallo, P. S. Meta-analysis–A mixed metaphor? *American Psychologist,* 1978, *33,* 515–517.

Gibbons, J. D., Olkin, I. & Sobel, M. *Selecting and Ordering Populations: A New Statistical Methodology.* New York: Wiley, 1977.

Glass, G. Primary, secondary and meta-analysis of research. *Educational Research,* 1976, *5,* 3–8.

Glass, G. Integrating findings: The meta-analysis of research. *Review of Research in Education,* 1977, *5,* 351–379.

Glass, G. Summarizing effect sizes. In R. Rosenthal (Ed.), *New Directions for Methodology of Social and Behavioral Science: Quantitative Assessment of Research Domains.* San Francisco: Jossey-Bass, 1980.

Glass, G., McGaw, B., & Smith, M. L. *Meta-analysis in Social Research.* Beverly Hills: Sage, 1981.

Glass, G., & Smith, M. L. Meta-analysis of research on the relationship of class-size and achievement. *Educational Evaluation and Policy Analysis,* 1979, *1,* 2–16.

Gottfredson, S. Evaluating psychological research reports. *American Psychologist,* 1978, *33,* 920–934.

Greenwald, A. G. Consequences of prejudice against the null hypothesis. *Psychological Bulletin,* 1975, *82,* 1–20.

Hall, J. A. Gender differences in nonverbal communication skills. In R. Rosenthal (Ed.), *New Directions for Methodology of Social and Behavioral Science: Quantitative Assessment of Research Domains.* San Francisco: Jossey-Bass, 1980.

Hedges, L. V., & Olkin, I. Clustering estimates of effect magnitude from independent studies. *Psychological Bulletin,* 1983, *93,* 563–573.

Hunter, J. E., Schmidt, F. L., & Jackson, G. *Meta-analysis: Accumulating Results Across Studies.* Beverly Hills: Sage, 1982.

Israel, M. A meta-analysis of the Los Angeles Unified School District's efforts in achieving its goals of desegregation: Improved academic achievement and race relations. *Dissertation Abstracts International,* 1981 (May), *41(11-A),* 4651–4652.

Kavale, K. A. The efficacy of student drug treatment for hyperactivity: A meta-analysis. *Journal of Learning Disabilities,* 1982, *15,* 280–289.

Kerlinger, F. N., & Pedhauzer, E. J. *Multiple Regression in Behavioral Research.* New York: Holt, Rinehart and Winston, 1973.

Kirk, R. E. *Experimental Design: Procedures for the Behavioral Sciences.* Monterey: Brooks/Cole, 1982.

Kulik, C. C., & Kulik, J. A. Research synthesis on ability grouping. *Educational Leadership,* 1982, *39,* 619–621.

Leviton, L. C., & Cook, T. D. What differentiates meta-analysis from other forms of review. *Journal of Personality,* 1981, *49,* 231–236.

Light, R. J. Capitalizing on variation: How conflicting research findings can be helpful for policy. *Educational Researcher,* 1979, *8,* 3–8.

Madron, T. W. Multiple regression for the TRS-80. *Byte,* 1981 (October), 430–447.

Marascuilo, L. A., & Levin, J. R. *Multivariate Statistics in the Social Sciences: A Researcher's Guide*. Monterey: Brooks/Cole, 1983.

Mintz, J. Integrating research evidence: A commentary on meta-analysis. *Journal of Consulting and Clinical Psychology*, 1983, *51*, 71–75.

Noonan, L. *Basic BASIC-English Dictionary*. Beaverton: Dilithium Press, 1982.

O'Grady, K. E. Probabilities and critical values for *z*, chi-square, *r*, *t* and *f*. *Behavior Research Methods and Instrumentation*, 1981, *13*, 55–56.

Olson, S. R. *Ideas and Data: The Process and Practice of Social Research*. Homewood: Dorsey Press, 1976.

Rachman, S. J., & Wilson, G. T. *The Effects of Psychological Therapy: Second Enlarged Edition*. New York: Pergamon Press, 1980.

Rosenthal, R. *Experimenter Effects in Behavioral Research*. (Rev. Ed., 1976) New York: Appleton-Century-Crofts, 1966.

Rosenthal, R. Experimenter expectancy and the reassuring nature of the null hypothesis decision procedure. *Psychological Bulletin Monograph Supplement*, 1968, *70*, 30–47.

Rosenthal, R. Interpersonal expectations. In R. Rosenthal & R. L. Rosnow (Eds.), *Artifact in Behavioral Research*. New York: Academic Press, 1969.

Rosenthal, R. Combining results of independent studies. *Psychological Bulletin*, 1978, *85*, 185–193.

Rosenthal, R. The "file drawer problem" and tolerance for null results. *Psychological Bulletin*, 1979, *86*, 638–641.

Rosenthal, R. *New Directions for Methodology of Social and Behavioral Science: Quantitative Assessment of Research Domains*. San Francisco: Jossey-Bass, 1980.

Rosenthal, R. Assessing the statistical and social importance of the effect of psychotherapy. *Journal of Consulting and Clinical Psychology*, 1983, *51*, 4–13. (a)

Rosenthal, R. Meta-analysis: Toward a more cumulative social science. In L. Bickman (Ed.), *Applied Social Psychology Annual* (vol. 4). Beverly Hills: Sage, 1983. (b)

Rosenthal, R. *Meta-analytic Procedures for Social and Behavioral Research*. Beverly Hills: Sage, in press.

Rosenthal, R., & Rosnow, R. L. *The Volunteer Subject*. New York: Wiley-Interscience, 1975.

Rosenthal, R., & Rosnow, R. L. *Essentials of Behavioral Research: Methods and Data Analysis*. New York: McGraw-Hill, 1984. (a)

Rosenthal, R., & Rosnow, R. L. *Essentials of Contrasts: Comparison Procedures in the Analysis of Variance*. Manuscript submitted for publication, 1984. (b)

Rosenthal, R., & Rubin, D. B. Pygmalion reaffirmed. In J. D. Elashoff and R. E. Snow (Eds.), *Pygmalion Reconsidered*. Worthington, OH: Charles A. Jonis, 1971.

Rosenthal, R., & Rubin, D. B. Interpersonal expectancy effects: The first 345 studies. *The Behavioral and Brain Sciences*, 1978, *3*, 377–415.

Rosenthal, R., & Rubin, D. B. Comparing significance levels of independent studies. *Psychological Bulletin*, 1979, *86*, 1165–1168.

Rosenthal, R., & Rubin, D. B. A simple, general purpose display of magnitude of experimental effect. *Journal of Educational Psychology*, 1982, *74*, 166–169. (a)

Rosenthal, R., & Rubin, D. B. Comparing effect sizes of independent studies. *Psychological Bulletin*, 1982, *92*, 500–504. (b)

Rosenthal, R., & Rubin, D. B. Further meta-analytic procedures for assessing cognitive gender differences. *Journal of Educational Psychology*, 1982, *74*, 708–712. (c)

Schmidt, F. L. *Moderator research and the law of small numbers*. Paper presented at the Conference of Moderator Research, University of Maryland, March, 1977.

Smart, R. G. The importance of negative results in psychologial research. *Canadian Psychologist*, 1964, *52*, 225–232.

Smith, M. L., & Glass, G. Meta-analysis of psychotherapy outcome studies. *American Psychologist*, 1977, *32*, 752–760.

Smith, M. L., Glass, G., & Miller, T. I. *The Benefits of Psychotherapy*. Baltimore: The Johns Hopkins University Press, 1980.

Snedecor, G. W., & Cochran, W. G. *Statistical Methods*. (6th Edition; 7th Edition). Ames: Iowa State University Press, 1967; 1980.

Sohn, D. Critique of Cooper's meta-analytic assessment of the findings on sex differences in conformity behavior. *Journal of Personality and Social Psychology*, 1980, *34*, 1215–1221.

Sokol, R. R., & Rohlf, F. J. *Biometry*. San Francisco: W. H. Freeman, 1969.

Sterling, T. D. Publication decisions and their possible effects on interfaces drawn from tests of significance—or vice versa. *Journal of the American Statistical Association*, 1959, *54*, 30–34.

Stock, W. A., Okun, M. A., Haring, M. J., Miller, W., Kinney, C., & Ceurvorst, R. W. Rigor in data synthesis: A case study of reliability in meta-analysis. *Educational Researcher*, 1982, *11*, 10–14.

Stouffer, S. A. *The American Soldier (vol. 1: Adjustment During Army Life)*. Princeton: Princeton University Press, 1949.

Strube, M. J., & Garcia, J. E. A meta-analytic investigation of Fiedler's contingency model of leadership effectiveness. *Psychological Bulletin*, 1981, *90*, 307–321.

Strube, M. J., & Hartmann, D. P. A critical appraisal of meta-analysis. *British Journal of Clinical Psychology*, 1982, *21*, 129–139.

Strube, M. J., & Hartmann, D. P. Meta-analysis: Techniques, applications and functions. *Journal of Consulting and Clinical Psychology*, 1983, *51*, 14–27.

Wallace, D. L. Bounds on normal approximations to student's and the chi-square distributions. *Annals of Mathematical Statistics*, 1959, *30*, 1121–1130.

Webb, E., Campbell, D., Schwarz, R., & Sechrest, L. *Unobtrusive Measures: Nonreactive Research in the Social Sciences*. Chicago: Rand-McNally, 1966.

White, K. R. The relation between socioeconomic status and academic achievement. *Psychological Bulletin*, 1982, *91*, 461–481.

Williams, P. A. The impact of leisure-time television on school learning: A research synthesis. *American Educational Research Journal*, 1982, *17*, 19–50.

Willson, V. L., & Putnam, R. R. A meta-analysis of protest sensitization effects in experimental design. *American Educational Research Journal*, 1982, *19*, 249–258.

Wilson, G. T., & Rachman, S. J. Meta-analysis and the evaluation of psychotherapy outcome: Limitations and liabilities. *Journal of Consulting and Clinical Psychology*, 1983, *51*, 54–64.

Zuckerman, M., DePaulo, B. M., & Rosenthal, R. Verbal and nonverbal communication of deception. In L. Berkowitz (Ed.), *Advances in Experimental Social Psychology* (vol. 14). New York: Academic Press, 1981.

Author Index

Subject Index